Long-Life Design and Test Technology of Typical Aircraft Structures

Jun Liu · Zhufeng Yue · Xiaoliang Geng
Shifeng Wen · Wuzhu Yan

Long-Life Design and Test Technology of Typical Aircraft Structures

Jun Liu
School of Mechanics, Civil Engineering and Architecture
Northwestern Polytechnical University
Xi'an, Shaanxi
China

Zhufeng Yue
Northwestern Polytechnical University
Xi'an, Shaanxi
China

Xiaoliang Geng
Northwestern Polytechnical University
Xi'an, Shaanxi
China

Shifeng Wen
Northwestern Polytechnical University
Xi'an, Shaanxi
China

Wuzhu Yan
Northwestern Polytechnical University
Xi'an, Shaanxi
China

ISBN 978-981-13-4140-3 ISBN 978-981-10-8399-0 (eBook)
https://doi.org/10.1007/978-981-10-8399-0

Jointly published with National Defense Industry Press, Beijing, China

The print edition is not for sale in China Mainland. Customers from China Mainland please order the print book from: National Defense Industry Press.

Printed on acid-free paper

This Springer imprint is published by the registered company Springer Nature Singapore Pte Ltd.
part of Springer Nature
The registered company address is: 152 Beach Road, #21-01/04 Gateway East, Singapore 189721, Singapore

Preface

Aviation technology is moving forward at an unprecedented speed. Aircraft structures tend to be large in scale, complex, lightweight, and sophisticated. However, the service environment of aircraft structures is getting worse. Its development increasingly focuses on improving reliability and reducing costs, that is, improving safety and extending material life under the premise of meeting structural strength requirements. Structural anti-fatigue design and manufacturing has become the key issue of aircraft reliability and economy. With the launch of large passenger aircraft projects and transport aircraft projects in China, typical structural designs of aircraft, and their related tests, have encountered new technical problems. Solutions to these key technological issues will be helpful to promote domestic aircraft design and testing technologies, as well as the comprehensive performance of aircraft, which has become particularly important.

Under the support of projects by the National Defense Science and Technology, the localization project of materials, the National Natural Science Foundation of China (50805118, 50375124, and 11102163), the Aviation Science Fund (2013ZA53010), and the basic research fund of the Northwestern Polytechnical University (JC20110260), our research group has achieved a lot in in terms of the exploration and the practice of anti-fatigue design, analysis, and related experimental verification of typical aircraft structures in recent years, which represents the basis of this book. The book's content considers practical engineering issues with high engineering application values. Part of the results and technologies have been successfully applied to different models of aircrafts in their structural design, analysis and test. The book is divided into six chapters.

Chapter 1 introduce the aim and main content of the book. Chapter 2 provides an overview of the basic concepts and knowledge often used in anti-fatigue design and analysis of aircraft structures. Chapter 3 analyzes the effect of the surface quality of fastener holes, such as surface defects and processing quality, on fatigue life. Chapter 4 examines the effects of fatigue resistance techniques, such as cold expansion, impression, and strengthening by hammering, on the residual stress distribution and fatigue life of fastening holes. Chapter 5 discusses the effects of different parameters of shot peening, such as shot materials, surface roughness, and target properties, on the residual stress field and lifespan of plates. Chapter 6 focuses on three types of typical connections for aircraft: single shear lap joints, double shear joint and reverse double dog bone joint in terms of nail load and lifespan assessment. Chapter 7 analyzes fatigue tests and lifespan assessments of typical wing box structures, and verifies the accuracy of anti-fatigue analysis technologies used on such structures to determine their anti-fatigue properties.

In the relevant pre-research and project completion process, Dr. Shao Xiaojun, doctoral students Kang Jianxiong and Yuan Xin, master students Liu Yongjun, Zhang Gang, Xu Honglu, Wang Xiaosen, Yang Shichao, Zhang Zhiguo, and Wang Xing, along with Mr. Yao Shile, Mr. Wei Xiaoming, and Mr. Han Zhe, engineers from the advanced materials and structural testing center of Northwestern Polytechnic University, undertook certain tasks. Some parts of the book adopt or refer to their relevant papers or work reports, and we are exceedingly grateful for their help. This book has also cited research results from both domestic and foreign experts and scholars. However, we may not always indicate original sources and apologize to the reader when this is the case.

The completion of this book has also been strongly supported by the National Defense Science and Industry Bureau, China Aviation Industry Corporation's related plants and other units, as well as relevant leadership, engineering, and technical personnel. We would like to express our sincere appreciation to all of them.

This book's content is characterized by systematic, comprehensive, and high engineering practicality. At the same time, it reflects both domestic and foreign research results in this field in recent years and as such could be used as a reference for relevant professional researchers and engineering and technical personnel.

We warmly welcome your criticism and corrections.

Xi'an, China

Jun Liu
Zhufeng Yue
Xiaoliang Geng
Shifeng Wen
Wuzhu Yan

Contents

Brief Introduction

Long-life design and test technologies of typical structures of aircraft were researched in this book. Typical structures covered fastening holes, shot peened plates, different types of joints, and wing box.

From this book, readers can learn the process of anti-fatigue design, manufacture, analysis, and test verification technologies and truly understand the essence of anti-fatigue technology. This book can be used as a reference book for researchers and technicians majoring in aircraft design, materials, and other related realms.

Chapter 1
Introduction

Today, international aviation science and technology is developing at an unprecedented rate. Aircraft structures tend to large in scale, complex, lightweight, and sophisticated. Meanwhile, the servicing conditions of mechanical parts and components are getting worse, something which greatly increases the possibility of failure due to fatigue. The key bearing components in a structure need to be capable of dealing with high loads, have long lifespans, have high reliability, and be able to deal with harsh environments, whilst also being able to manage other conditions. At the same time, people are paying more and more attention to improving the reliability and economy of aircraft use, that is, to achieve aircraft safety and increased aircraft lifespans under the premise of meeting structural strength requirements, something which has stimulated the rapid development of fatigue research. Anti-fatigue design and manufacturing has become the key issue facing the reliability of aircraft structures.

Aircraft structural design methods have gone through several stages: static strength design, safety life design, damage tolerance design, and durability design [1]. In conditions found in early aircraft using static strength design methods, the static strength reserve was large, and aircraft stresses very low, so fatigue problems were covered. Since the 1950s, the aviation industry has experienced comprehensive development. In order to achieve high speeds and good flight performances, the aircraft must be as light as possible, with its static strength reserve becoming lower, however, fatigue problems have arisen. In 1954, the British jet plane "comet-I" collapsed twice due to fatigue cracks at the edge of its rivet connection areas. Consequently, people began to focus on aircraft fatigue strength design [2]. In the late 1960s, F-111 aircraft crashes, and other fatigue problems, provided the basis for the B-1 bomber to use fracture mechanics concepts in the 1970s. The United States Air Force in 1975 issued MIL-STD-1530A outlining that damage tolerance was required in new military aircrafts.

An aircraft's damage tolerance design is mainly aimed at its flight safety structure. The key issue is that the structure must have sufficient residual strength when in service. However, the damage tolerance design principle does not take

J. Liu et al., *Long-Life Design and Test Technology of Typical Aircraft Structures*,
https://doi.org/10.1007/978-981-10-8399-0_1

crack initiation into account in an aircraft's lifespan estimate. Instead, it assume a certain length of crack as equivalent initial crack length of structure, which is not real crack length, since it is random variable depending on the manufacturing process and capacity of detection. So there are some limitations in to this method. Aircraft durability design, for an aircraft's main and secondary bearing structures, is an important way to ensure its long life, its high reliability, its high economy, and continuous combat effectiveness. Modern aircraft structural design has entered the stage of durability/damage tolerance design which centers on structural integrity. The main criteria of modern aircraft structural design are the combination of durability design and damage tolerance design, representing the most advanced design method so far [3, 4].

With the launch of large-scale passenger aircraft projects and transport projects in China, new technical problems have emerged which are key to the structural design of aircraft and related tests. Solution of these key technology issues will help improve the design of aircraft structures and the comprehensive performance of the aircraft, as well as helping to develop relevant test technologies.

Fatigue cracking of fastening holes is one of the most common causes of damage to in-service aircraft. Therefore, the major topics of aircraft design are how to minimize the stress concentration impact of fastening holes, improve anti-fatigue performance of aircraft structures, extend service life, and ensure the reliability and safety of aircraft structures through aircraft design, material selection, and manufacturing.

In engineering practice, surface hardening of fastening holes and panels is usually used to extend an aircraft's life. Strengthening treatments refer to changing the structure and stress distribution of the material without changing the material itself, or its cross-sectional dimensions, with the purpose of improving structural fatigue life. Fastening hole surface treatment technologies are mainly linked to, but not limited to, cold extrusion, interference fit, and embossing. For plates, surface treatment technologies include shot peening strengthening and forming, laser strengthening, rolling, as well as other technologies. The most important problems being faced are how to optimize process parameters and maximize the anti-fatigue performance of components as much as possible.

There are a large number of connectors in an aircraft's structure, and the discontinuity of the structure and the complex forms of force acting on these component structures make these connectors susceptible to fatigue damage. The material used for the connector, the connecting method of screwing or riveting, the material used for the nails, the positions of the nail holes, the lap joint form of the plate, and so on, all have a great influence on the nail load and stress distribution across the connector, hence affecting the structural anti-fatigue performance.

At present, aircraft design has entered a stage of durability and damage tolerance design. For the main bearing components, such as the fuselage frame and wing box, the main issues are how to determine the original fatigue quality of the weak parts using fatigue tests, selecting an effective damage tolerance assessment method, setting a structural detecting cycle using appropriate maintenance technology, and determining the economic life of components by consulting the durability design method.

In terms of lifespan prediction technology, based on different fatigue mechanisms and assumptions, many domestic and foreign scholars have proposed a variety of fatigue lifespan prediction methods or theories. These include the nominal stress method in uniaxial fatigue, the local stress–strain method, linear elastic fracture mechanics, the energy method in multiaxial fatigue, the equivalent stress–strain method, and the critical plane method. Since these theories and methods are drawn based on certain assumptions and conditions, and adding the scatter of fatigue test, there are some difference between prediction results and test results. It is important to choose effectively the fatigue prediction technology according to the bearing state and stress–strain distribution of the structure, obtaining a conservative solution which is as accurate as possible in order to provide reference for aircraft design, material selection, and manufacturing processes.

Modern aircraft generally use a wing structure design incorporating fuel tanks, in which the closed wing box is used as a storage space. As the overall fuel tank plays a role in both structural load and fuel storage, the design has to consider not only the structural strength and stiffness requirements, but also the sealing requirements. The corrosion resistance of the whole tank, and the sealing performance at different temperatures, are the main risks affecting the safety and reliability of the aircraft. If the oil tank has a serious leak or oil spill, the aircraft will be endangered. It is necessary to reasonably determine a test scheme and corrosion resistance and make sure all sealing meets the design requirements.

For years, our research team has completed work on aircraft structure anti-fatigue technology and related test verification methods. This book is a summary of the research work by the author and research group, which consists of anti-fatigue design, analysis, and test technologies for key structural aircraft parts, based on existing problems in structural design and practical engineering.

References

1. *United States Durability Design Handbook*. 1991. Systems Engineering of AFFD.
2. Suresh, S., and Zhongguang Wang, the translator, *Material Fatigue Second Version*. National Defense Industry Press.
3. Zhentong Gao and Junjiang Xiong, 2001. *Fatigue Reliability*, Beihang University Press.
4. *Analysis and Design Guide of Military Aircraft Structure Durability/Damage Tolerance, China Institute of Aircraft Strength*, 2005.

Chapter 2
Outline of Fatigue and Fracture Mechanics

Fatigue is a branch of research comprised of many disciplines, requiring a broad knowledge not only of elastic mechanics, plastic mechanics, fracture mechanics, vibration, and stress analysis, but also mathematics, metallurgy, machinery, metal physics, and other inextricably linked disciplines. It includes fields of research in the microscopic range, macroscopic modelling, mechanics, and theory and analysis of experimental science. This chapter briefly introduces some of the basic knowledge that is often used in anti-fatigue design and analysis of aircraft structures.

2.1 Basic Conception of Fatigue

2.1.1 Definition of Fatigue and Its Damage Property

In 1964, the Geneva International Standard Organization (ISO) defined fatigue as a property change of a metal material under repeated stress or strain after undergoing the General Principles of Metal Fatigue Test. American Society for Testing and Materials's definition of fatigue is the process of gradual, permanent, partial change in a material structure, which happens over the course of changes of one or several points in a material caused by stress and strain developing a crack or fracture [1].

Generally, the process of fracturing is divided into three stages: crack nucleation, crack extending, and fracture [2]. Crack nucleation lifespan is represented by the cycles taken by a material or structure, after bearing alternating loads, to form a crack of a given length, a_0. Usually, a_0 is the minimum length detected by present-day instruments. Crack extending lifespan represents the cycles for the development of the initial crack to the final fracture. Total fatigue lifespan is the combined crack nucleation lifespan and crack extending lifespan.

Usually, fatigue damage demonstrates several of the following features [3].

J. Liu et al., *Long-Life Design and Test Technology of Typical Aircraft Structures*,
https://doi.org/10.1007/978-981-10-8399-0_2

1) Damage may happen in conditions when alternating stress is much less than the material's static strength limit.
2) The appearance of fatigue damage must experience some stress and strain cycles.
3) Fatigue damage is dangerous because it is often invisible during maintenance and repair because parts or samples of a structure seldom show macroscopic plastic deformation but instead show something like brittle failure.
4) There are two obvious areas in a fatigue fracture, one is a smooth area and the other is final failure area.

2.1.2 Alternating Stress

Alternating stress is a stress that changes periodically, which is marked by S. A periodic change of stress is called a stress cycle as shown in graph (Fig. 2.1).

In stress cycle, $s_{\max}$ represents the maximum stress, $s_{\min}$ the minimum stress, s_m the average stress, R the stress ratio, and s_a the stress amplitude.

$$s_a = \frac{s_{\max} - s_{\min}}{2} \tag{2.1}$$

$$R = \frac{s_{\min}}{s_{\max}} \tag{2.2}$$

$$s_m = \frac{s_{\max} + s_{\min}}{2} \tag{2.3}$$

The five parameters of one stress have the relationship as shown in (2.1), (2.2), and (2.3), so there are only two independent parameters. If given any two of the five parameters, the rest will be calculated. So to describe a fatigue load spectrum with constant amplitude, two of the five parameters are enough.

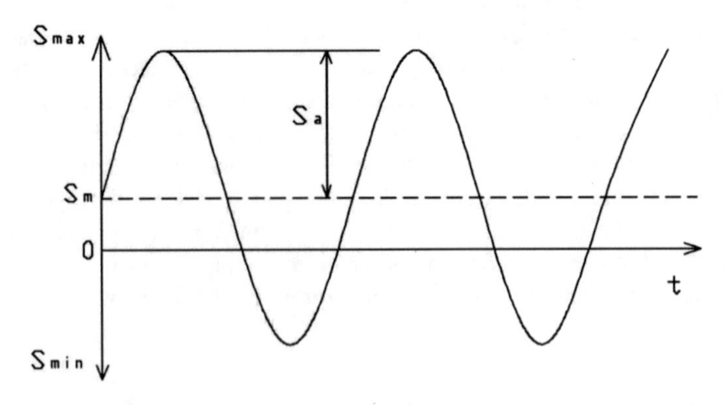

Fig. 2.1 Alternating stress

Besides, there are several special cases that often happen which are as follows:

1. Symmetrical cycle, that is, if $s_m = 0$, $s_{\max} = -s_{\min}$, and $R = -1$
2. Pulsating cycle, that is, if $s_{\min} = 0$, $R = 0$
3. Stress under statics bearing, that is if $s_a = 0$, $s_{\max} = s_{\min} = s_m$ and R = 1

2.1.3 Curve S-N

Under an unchangeable stress ratio *R*, apply alternating bearing on a set of samples with different $s_{\max}$ until fractures, record each fracture cycle of the samples, and make $s_{\max}$ as ordinate and fracture cycles *N* as abscissa to make a curve, then we get the material *S-N* curve under a given stress ratio *R*.

The whole *S-N* curve contains three parts they are low cycle fatigue (LCF), high cycle fatigue (HCF), and secondary fatigue (SF) as shown in graph (Fig. 2.2). In logarithmic coordinate systems, *S-N* curve is similar to a straight line on HCF region.

So far, there are three experiential equations to express the approximately *S-N* curve

1. Power function

$$S_a^m N = C \tag{2.4}$$

In the above equality, *m* and *c* are constant in connection with material property, sample forms, and loading ways and are determined by test.

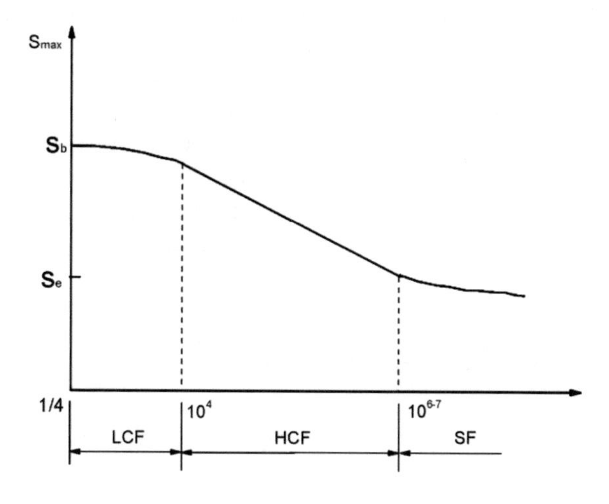

Fig. 2.2 Typical *S-N* curve

The equality (2.4) describes the power function relationship between the stress amplitude S_a and life N, on the condition that the stress ratio R or average stress S_m is given. Taking logarithm on both sides of the equality, we get

$$\lg N = \lg C - m \lg S_a \tag{2.5}$$

Therefore, the power function show that it is linear relationship between $\lg S_a$ and $\lg N$ in double logarithm coordinate system.

2. Exponential function

$$\mathrm{e}^{mS_{\max}} N = C \tag{2.6}$$

m and C are two material constants from test. Equality (2.6) shows the exponential relationship between the maximum stress $S_{\max}$ and life N with a given stress ratio R or an average stress $S_{m.}$ Taking logarithm in both sides of the equality (2.6), we get

$$\lg N = \lg C - mS_{\max} \lg e \tag{2.7}$$

Therefore, exponential function show that $S_{\max}$ and $\lg N$ has linear relationship in single logarithm coordinate system.

3. Power function with the three parameters

$$S_{\max} = S_{\infty}\left(1 + \frac{A}{N^{\alpha}}\right) \tag{2.8}$$

here, A, α and S_{∞} are material constant, and S_{∞} equals to the maximum stress when $N \to \infty$, which can approximately represent stress limit.

In addition, there are horizontal progressive lines in S-N curves of steel and cast iron proved by tests, which means if the stress is more than a value of S, the sample will fracture after some cycles, but if the stress is less than the value, the sample will never fracture even after infinite cycles. The value of S just mention is called material fatigue limit under a given stress ratio R. Fatigue limit in manual is usually the one when $R = -1$, which is identified by S_{-1}

2.1.4 Equal-Life Curve

Material S-N curve will change if the stress ratio R is changed. If given several values of stress ratio of a material, we will get correspondingly different S-N curves which are called curve family. Based on each stress ratio R and the related $S_{\max}$, we can get S_a and S_m, and equal-life curves come from the connection of equal-life points on the plane S_a-S_m

Today, the commonly used experiential equal-life curve expressions are as follows:

1. Goodman linear equation

$$\frac{S_a}{S_{-1}} + \frac{S_m}{\sigma_b} = 1 \tag{2.9}$$

σ_b is the material tensile strength limit and S_{-1} is the fatigue limit under symmetric cyclic loading.

2. Gerber parabolic equation

$$\frac{S_a}{S_{-1}} + \left(\frac{S_m}{\sigma_b}\right)^2 = 1 \tag{2.10}$$

3. Soderberg linear equation

$$\frac{S_a}{S_{-1}} + \frac{S_m}{\sigma_s} = 1 \tag{2.11}$$

σ_s is the material tensile yield limit.

4. Cepehceh discount equation

$$\frac{S_a}{S_{-1}} + \left[\left(\frac{2}{S_0} - \frac{1}{S_{-1}}\right)S_m\right] = 1 \tag{2.12}$$

S_0 is the material fatigue limit under successive cyclic loading.

5. Bagci quartic equation

$$\frac{S_a}{S_{-1}} + \left(\frac{S_m}{\sigma_b}\right)^4 = 1 \tag{2.13}$$

2.1.5 *Stress Fatigue and Strain Fatigue*

Fatigue is divided into two types, one is high cycle fatigue (HCF) or stress fatigue, the other is low cycle fatigue (LCF) or strain fatigue, according to different levels of stress or fatigue cycles. Differences between them are as shown below in diagram (Table 2.1).

Table 2.1 Differences between HCF and LCF

Types of fatigue	HCF	LCF
Definition	Fatigue with fracture cycles more than 10^4	Fatigue with fracture cycles less than 10^4
Stress	Bellow the elastic limit	Above the elastic limit
Plastic deformation	Without obvious macroscopic plastic deformation	With obvious macroscopic plastic deformation
Stress–strain relationship	Linear relationship	Nonlinear relationship
Controlling parameter	Stress	Strain

Usually, stress changes a little but strain changes a lot when the material is in plasticity, under which strain controlling is more reasonable. Therefore, if it is LCF, life is calculated out by ε–N curve (ε is strain and N is fatigue life).

In all of the ε–N curves, Manson–Coffin equation is the most commonly used one with the expression as shown below.

$$\left(\frac{\Delta\varepsilon}{2}\right)_{\text{total}} = \frac{\sigma_f^{'}}{E}\left(2N_f\right)^b + \varepsilon_f^{'}\left(2N_f\right)^c \tag{2.14}$$

$\sigma_{\rm f}^{'}$ is fatigue strength coefficient and b is fatigue strength exponent. $\varepsilon_{\rm f}^{'}$ is fatigue ductility coefficient and c is fatigue ductility exponent.

2.1.6 *Linear Cumulative Damage Theory*

Linear cumulative fatigue damage theory is that fatigue damage can be cumulated linearly under cycle loadings. When these stresses, which are independent and have nothing to do with each other, are accumulated to a certain value, fatigue rupture of samples or structure will appear.

The Miner linear cumulative damage theory is the most famous one in all of them. Its main idea is that for a load sequence consist of m blocks of load cycles with respectively constant stress amplitude, if n_i represents the fracture cycles of the ith loading block under the constant stress amplitude $\Delta\delta_i$, and n_{fi} represents fracture cycles under $\Delta\delta_i$, damage is the one as follows.

$$D = \sum_{i=1}^{m} n_i/n_{fi} \tag{2.15}$$

Here, failure occurs if $D = 1$. Linear cumulative damage theory is used to quickly predict the fatigue life in initial design phase for its simple form.

2.2 The Main Factors that Affect the Structural Fatigue Performance

2.2.1 *Effects of Load Spectrum*

The main load factors that affect the fatigue strength of the material include load type, loading frequency, average stress and load duration.

In general, the effect of frequency on the fatigue crack growth rate $\mathrm{d}a/\mathrm{d}N$ is much less affected than the stress ratio R. In noncorrosive environment at room temperature, the effect of frequency changes from 0.1 to 100 Hz on $\mathrm{d}a/\mathrm{d}N$ can hardly be considered. Stress cycle waveform (sine wave, triangular wave, rectangular wave, etc.) is less important. When the fatigue load is symmetrical cyclic load, the average stress is zero. When the average stress is greater than zero, the fatigue life increases with the increase in the average stress. So the load continues to make the fatigue strength or fatigue life to decrease, but not significantly.

2.2.2 *Effects of Stress Concentration*

Stress and strain will increase in the discontinuous location of structural such as notch or hole edge. This phenomenon is called stress concentration. In engineering practice, stress concentration is inevitable, and it can only be alleviated and weakened. Under the cyclic load, when the nominal stress is less than the material yield stress, the stress concentration area has often entered the plasticity. Therefore, the stress concentration is the weakness of structural fatigue. The severity of the stress concentration can be described by the theoretical stress concentration factor.

$$K_T = \frac{\text{Maximum local elastic stress } \sigma_{\max}}{\text{Nominal stress } \sigma_0} \tag{2.16}$$

For some complex structures, you can also use finite element method to find K_T.

2.2.3 *Effects of Size*

It was found that the fatigue strength of large-sized specimens was often lower than that of small-sized specimens in fatigue strength tests. This is because the large-sized specimen contains more defects than the small-sized specimen, and the stress of large-sized specimen fatigue damage area is more serious, and more easily destroyed than the stress of the specimen with small size.

2.2.4 *Effects of Surface Roughness and Residual Stress*

In general, fatigue cracks originate from the surface. According to the theory of crystal plasticity (microstructure), the surface roughness is equivalent to surface intrusion and extrusion, thereby shortening the fatigue crack initiation life. According to the theory of fracture mechanics (macroscopic structure), the bigger the surface roughness value is, the deeper the groove of the surface will be. The smaller the radius of the bottom is, the more serious the stress concentration will be, and the worse the ability of fatigue resistance will be. So the increase in surface roughness will reduce the fatigue strength of the parts.

In order to improve the fatigue performance, in addition to decreasing the surface roughness, another methods which introducing compressive stress in the surface of component are usually used to achieve the purpose of improving fatigue life. Residual compressive stress fatigue strengthening mechanism was shown in Fig. 2.3, it can be seen that the surface residual compressive stress results in a decrease in the average stress of the cyclic load, which achieves fatigue strengthening

In engineering, some important parts are treated by shot peening, cold expansion, and other surface deformation strengthening process, thereby introducing residual compressive stress in the surface of the material, causing surface layer strengthening, and improving the fatigue properties of the material. It should also be noted that fatigue damage is mainly induced by tensile stress. Residual stress is a self-balancing system. Large cold deformation may induce micro-cracks on the surface, reducing the fatigue performance of parts.

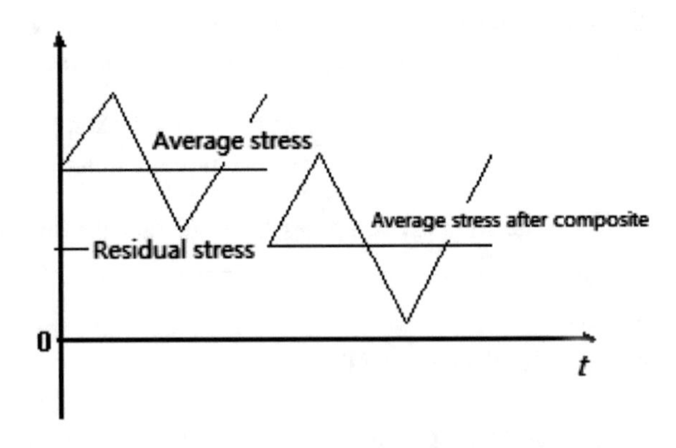

Fig. 2.3 Residual stress decreases the average cyclic stress

2.3 Mechanism of Fatigue Failure, Crack Propagation, and Fracture Analysis of Metals

2.3.1 Mechanism of Fatigue Failure

In general, the fatigue damage of metal can be divided into three stages: fatigue crack initiation, fatigue crack propagation, and instability fracture.

1. **Fatigue crack initiation**

The initiation of fatigue cracks is caused by local plastic strain concentration. There are three common modes of crack initiation: slip band cracking, grain boundary and twin crystal boundary cracking, and interface cracking between the inclusions or second phase and substrate. Slip band cracking is not only the most common way of fatigue crack initiation but also the most basic of the three initiation methods. The reason is that plastic deformation and formation of slip band have exist before grain boundary cracking and interface cracking occur. Thus, the ductile deformation and the formation of the slip band are not only the basis for the slip band but also a prerequisite for the other two initiation modes.

2. **Fatigue crack propagation**

Fatigue crack propagation can be divided into two stages, I and II. In the first stage, the micro-cracks first grow along the maximum activity of the shear stress and have certain crystallographic properties. Under uniaxial stress, crack grows along the sliding surface at an angle of approximately 45° with stress direction. Crack propagation rate is very slow in stage I, generally lower than 3×10^{-7}mm/cycle.

When the micro-cracks are extended to the depth of a grain or two or three grains, the direction of propagation of the crack is gradually shifted from the direction of the angle of 45° at the beginning to the direction perpendicular to the tensile stress. From then on, crack propagation is in the second stage. In the second stage of crack propagation, only one main crack is left, and the crack propagation rate is usually at $\mathrm{d}a/\mathrm{d}N = 3 \times (10^{-7} - 10^{-2})$ mm/cycle. Crack propagation is subjected to shear stress control at stage I, and crack propagation is controlled by normal stress at stage II.

3. **Fracture**

Fracture is the final stage of fatigue damage. It occur suddenly and it is different with previous two stages. But the whole process of fatigue, which is still progressive, is caused by the gradual accumulation of damage. The fracture is the result of the crack extending to the critical dimension, and the stress intensity factor of the crack tip reaching the critical value.

2.3.2 Fatigue Crack Propagating Theory

Fatigue crack propagation rate $\mathrm{d}a/\mathrm{d}N$ is the amount of crack propagation in the alternating load per cycle, in the process of fatigue crack propagation, $\mathrm{d}a/\mathrm{d}N$ is changing Constantly, each instantaneous $\mathrm{d}a/\mathrm{d}N$ is the slope of the a–N curve at that point of the crack length, a varies with the number of alternating load cycles N. The crack propagation rate $\mathrm{d}a/\mathrm{d}N$ is controlled by the alternating stress field of the crack front, which is mainly the range of the stress intensity factor of the crack tip ΔK and the stress ratio of the alternating load R. The linear elastic fracture mechanics assumes that the change in the stress is invariant under the alternating load. The variation of $\mathrm{d}a/\mathrm{d}N$ with ΔK is shown in the double logarithmic coordinate system as shown in Fig. 2.4.

The curve of $\mathrm{d}a/\mathrm{d}N - \Delta K$ is divided into three stages: the threshold region I, the stable crack expansion region II, and unstable crack growth region III. In the first stage, the crack front edge stress field strength factor ΔK is low and the crack propagation rate $\mathrm{d}a/\mathrm{d}N$ is low. The vertical progressive line of stage I $\Delta K = \Delta K_{\mathrm{th}}$ is called the crack propagation threshold, when $\Delta K < \Delta K_{\mathrm{th}}$, the crack is almost no expansion. The second stage is the stable stage of fatigue crack propagation, and it is the main stage of determining fatigue life. It is generally believed that in the double logarithmic coordinate system, the relationship of $\mathrm{d}a/\mathrm{d}N$ and ΔK between this stage is a straight line, also known as linear crack extension. The vertical progressive line $\Delta K = \Delta K_c$ at stage III is the fracture toughness of the material. At stage III, since the maximum stress intensity factor tends to fracture toughness, crack rapidly extend. Stage III is known as unstable crack growth region and corresponding crack propagation life is small in whole crack growth process. Therefore, the main consideration of the crack expansion about the establishment of the equation to describe the crack growth rate is stage I, II [3, 4].

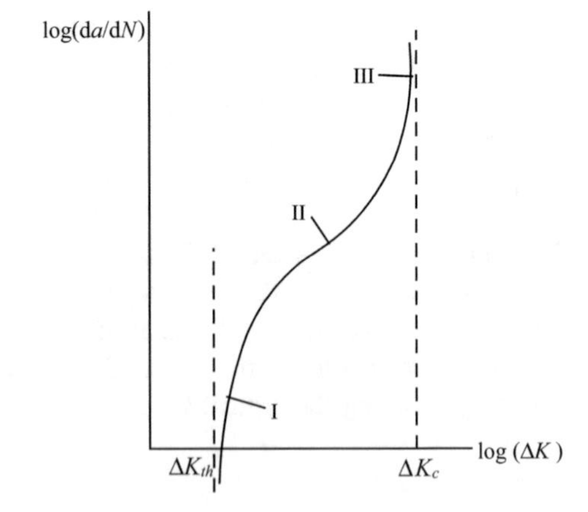

Fig. 2.4 $\mathrm{d}a/\mathrm{d}N - \Delta K$ curve shape

In general, the fatigue crack growth rate is a function of stress intensity factor ΔK, stress ratio R, material thickness, and environmental factors. The common equation that describes the rate of crack propagation is the Paris equation:

$$\mathrm{d}a/\mathrm{d}N = C(\Delta K)^n \tag{2.17}$$

In Eq. (2.17), C and n are material constants. The Paris equation has been widely used due to its simple form, and it can better describe the second stage of crack propagation.

2.3.3 Fractographic Analysis

A typical fatigue fracture usually presents three areas with different topography, a source of fatigue, a crack extension, and a transient crack zone. Figure 2.5 shows the macroscopic morphology of typical fatigue fracture. The most significant macroscopic morphology of fatigue fracture is that there is no obvious plastic deformation and can be divided into two distinct regions—the smooth fatigue zone and the rugged burst area [5].

1. **Fatigue source area**

Throughout the fracture, the area of fatigue source is the smallest. In general, fatigue cracks often occur on the surface, but for the existence of stress concentration of parts, fatigue cracks often initiate in the corner or rounded root. The source region is mostly semicircular or semielliptical in the fracture, because the expansion rate of the crack in the source area is slower, so it is more flat.

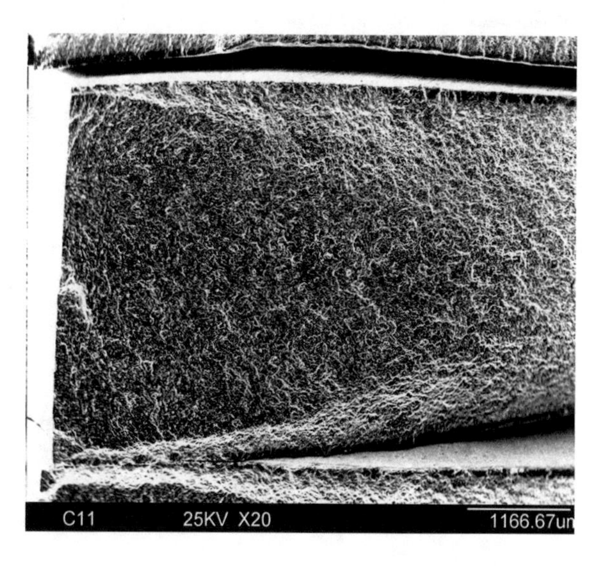

Fig. 2.5 Typical fatigue fracture

Fig. 2.6 Fatigue strip

2. **Fatigue crack propagation zone**

The area is the largest in the fracture. It locates between the fatigue source and transient crack zone. And it is not flat as fatigue source area. The microstructure of the crack extension is shown in Fig. 2.6, and the main microscopic feature of the fracture is that it has a fatigue band. These stripes are slightly curved, parallel to the same piece, with regular spacing, perpendicular to the direction of crack propagation. The fatigue strip is a unique fracture characteristic of the material under fatigue stripes. Each fatigue strip corresponds to a single load cycle, indicating the location of the crack front end of the cycle. In general, the number of fatigue bands is equal to the number of cycles. The spacing of the fatigue strip Δa is proportional to the stress intensity factor of the crack tip ΔK, and the higher the ΔK, the greater the Δa. According to the characteristic of the fatigue band corresponding to the load cycle, the expansion rate of the fatigue crack can be calculated, and the fatigue life of the part can be estimated.

3. **Transient crack zone**

Transient crack zone is coarse grain at usual because the contact time of this area and the environment medium is shorter, the section has a bright color, and the size of the instantaneous fault area depends mainly on the load and material properties.

2.4 Estimate Methods of Fatigue Life

2.4.1 Nominal Stress Approach

Nominal stress approach, the earliest way of fatigue life estimating, is a method of verifying fatigue strength or calculating fatigue life based on *S-N* curves of material or structure, compared with stress concentration factor and nominal stress of fatigue critical sites of material or structure, in accordance with cumulative fatigue damage theory.

Nominal stress approach makes an assumption that if values of stress concentration factor K_T as well as loading amplitudes are same for different components made of the same material, the lives of these components are same. In this approach, nominal stress and stress concentration factor are the controlling parameters.

The steps to estimate structure fatigue life by nominal stress approach are as follows:

1. Find out the fatigue critical site.
2. Calculate the nominal stress and stress concentration factor K_T of the fatigue critical site.
3. Determine the nominal stress amplitude of the fatigue critical site according to the loading amplitude.
4. Calculate the present stress concentration factor and *S-N* curve under the stress by interpolation method, and check *S-N* curve. What should be noticed is that before interpolation, we should take the log of test data, because *S-N* curve conforms well to quadratic curve in semi-logarithm coordinate system.
5. Work out fatigue life of the critical site by cumulative fatigue damage theory. There are two main deficiencies of the nominal stress approach. One is that it has not considered local plastic of the fracture foot, and the other is that it is quite difficult to determine the equivalent relationship between standard sample and structure, since the relationship has something to do with many factors, such as structure geometry, loading method, and material and size of the structure.

Because of the above defects, the ability of nominal stress approach to estimate fatigue crack forming life is low. Besides, it needs *S-N* curve under different stress rates R and stress concentration factors K_T to work, which will cost more time and money because more data of material is needed.

In the development of nominal stress method, there appear stress severity factor method (SSF), effective stress method, detail rated factor (DRF), and so on.

2.4.2 Local Stress–Strain Method

In practice, the material fatigue strength and life depend on the maximum local stress and strain of the strain concentration. Therefore, local stress–strain method, a new way to estimate fatigue life, was promoted on the basis of low fatigue cycle equation. In the new method, life is estimated according to the local stress–strain course of the critical site, with the local stress–strain amplitude of the critical site coming from nominal stress amplitude of component, by elastoplastic finite element or other ways, consulting cycle stress–strain curve of material.

Basic assumption of local stress–strain is that if the maximum stress–strain course of critical site made of the same material is the same with a smooth sample's stress–strain course, their fatigue lives are the same. In this method, the controlling parameter is local stress–strain.

The steps of local stress–strain method to estimate structure fatigue life are as follows.

1. Find out the fatigue critical site of the structure;
2. Work out the nominal stress amplitude of the critical site;
3. Calculate local stress-strain amplitude by elastoplastic finite element or other ways;
4. Find out the ε-N curve under the present stress–strain.
5. Work out fatigue life of the critical site by cumulative fatigue damage theory.

Therefore, σ-ε curve and ε-N_f curve of material are needed in the course of estimating fatigue crack forming life by the way of local stress–strain method.

Local stress–strain method has overcome the two main defects of nominal stress method, but it has its own disadvantage, that is point stress criterion. So, the effect of stress gradient of the crack foot and multiaxial stress cannot be considered.

Nominal stress method is used to estimate entire life, but the local stress–strain method is used for crack forming life. The latter always works together with fracture mechanics, which means estimating crack forming life first and then extending life by the way of fracture mechanics. The total life is the summation of the two lives before.

2.4.3 Multiaxial Fatigue Theory

Generally speaking, in engineering practice, loading of structure is complex, which means the related primary stresses are not proportional or do not change in a loading cycle. In early times, multiaxial issues were transformed into uniaxial issues according to static strength theory. Then, uniaxial fatigue theory was used to solve the multiaxial fatigue issues, which were viable when solving the multiaxial issues under proportional loading.

Actually, the key components of engineering structure and equipment mostly serve under disproportional multiaxial loading under which the fatigue is quite different from these in uniaxial fatigue loading or proportional multiaxial fatigue loading. Especially, under disproportional varying amplitude loading, simple cycle technology cannot be used, which is different from that under uniaxial loading. So, there are a lot of difficulties to estimate fatigue life alone by the way of typical uniaxial theory [4].

Multiaxial fatigue failure criteria are divided into three types according to state parameters. The first is based on equivalent stress–strain method, the second is based on energy method, and the third critical is based on plane method. However, the first one lacks physical significance and cannot reflect stress–strain history. Energy is scalar, so the second one cannot get the direction of crack propaganda. Among these failure criteria, critical plane method is widely used.

There are two steps in critical plane method. The stress–strain history of fatigue critical plane is calculated first, and then the stress–strain of critical plane is transformed into cumulative fatigue damage. Fatigue crack propaganda is controlled by two parameters, one is the maximum shearing strain and the other is normal strain of the plane where the maximum shearing strain locates. The failure criteria based on critical plane method are to simplify multiaxial stress to the equivalent uniaxial stress finally. So far, the mostly used models based on critical plane method are as follows [6]:

1. Smith–Watson–Topper model (SWT)

In HCF, strain range is in elastic range of material. Basquin's equation can be used to estimate life, which is effected by the average stress.

Basquin's equation:

$$\left(\frac{\Delta\varepsilon_e}{2}\right) = \frac{\sigma_f^{'} - \sigma_m}{E}\left(2N_f\right)^b \tag{2.18}$$

In the above equation, $\sigma_f^{'}$, σ_m, and b are fatigue strength coefficient, average stress, and index.

Strain range in LCF is in plastic range of material. Manson–Coffin equation can be used to calculate life without effect of average stress.

Manson–Coffin equation:

$$\left(\frac{\Delta\varepsilon_p}{2}\right) = \varepsilon_f^{'}\left(2N_f\right)^c \tag{2.19}$$

In the above equation, $\varepsilon_f^{'}$ and c are separately fatigue ductility coefficient and fatigue ductility exponent.

Total strain-life equation:

$$\left(\frac{\Delta\varepsilon}{2}\right)_{\text{total}} = \frac{\sigma_f^{'} - \sigma_m}{E}\left(2N_f\right)^b + \varepsilon_f^{'}\left(2N_f\right)^c \tag{2.20}$$

Smith–Watson–Topper theory's prime point is crack's appearance and extending is mainly decided by normal stress or normal strain under some loadings, which is based on critical plane method. Smith, Watson, and Topper propose a new simple parameter SWT, considering the maximum normal stain range on the plane and also the effect of the maximum stress. The equation is as follows:

$$\frac{\Delta\varepsilon_{\max}}{2}\sigma_{n.\max} = \frac{\left(\sigma_f\right)^2}{E}\left(2N_f\right)^{2b} + \varepsilon_f\sigma_f\left(2N_f\right)^{b+c} \tag{2.21}$$

In the above equation, b is fatigue strength exponent, C is fatigue ductility exponent, and σ_f is the fatigue strength coefficient, which can be treated as the true stress in fracturing under static stretching in simple calculation. ε_f is the fatigue ductility coefficient and can be used as the true strain in fracturing under static stretching in simple estimation. $\Delta\varepsilon_{\max}$ and $\sigma_{n.\max}$ are, respectively, the normal stain range on critical plane and the maximum normal stress.

2. Wang–Brown (WB) Model

Considering the effect of shearing strain and average stress on fatigue life, Wang and Brown find a new model for fatigue life estimation based on Manson–Coffin equation, as follows:

$$\frac{\Delta v_{\max}}{2} + S\frac{\Delta\varepsilon_n}{2} = A\frac{\sigma_f - 2\sigma_{n.\text{mean}}}{E}(2N_f)^b + B\varepsilon_f(2N_f)^c \tag{2.22}$$

In this equation, $A = (1+\mu_c) + (1-\mu_c)S$, $B = (1+\mu_p) + (1-\mu_p)S$, S = 0.3, μ_c is material Poisson ratio. μ_p is Poisson ratio in the condition of plasticity and the value of it can be 0.5. $\Delta v_{\max}$ and $\Delta\varepsilon_n$ are separately shearing strain range and normal strain range on critical plane. $\sigma_{n.\text{mean}}$ is the average normal stress on critical plane and other parameters are same with these in Eq. (2.21).

According to the above model, the way to determine value of critical plane is as follows, so is the process.

Considering the complexity of the way to estimate azimuth of critical plane under triaxial stress, here simplified two-dimensional model is used. Two-dimensional stress/strain conversion equation is as follows:

$$\sigma'_{11} = \frac{\sigma_{11} + \sigma_{22}}{2} + \frac{\sigma_{11} - \sigma_{22}}{2}\cos 2\theta_i + \tau_{12}\sin 2\theta_i \tag{2.23}$$

$$\varepsilon'_{11} = \frac{\varepsilon_{11} + \varepsilon_{22}}{2} + \frac{\varepsilon_{11} - \varepsilon_{22}}{2}\cos 2\theta_i + \varepsilon_{12}\sin 2\theta_i \tag{2.24}$$

$$\gamma'_{12} = \frac{\varepsilon_{11} - \varepsilon_{22}}{2}\sin 2\theta_i - \varepsilon_{12}\cos 2\theta_i \tag{2.25}$$

σ_{11}, σ_{22},and τ_{12} are normal stress and tangential stress before conversion.

ε_{11}, ε_{22}, and ε_{12} are normal strain and tangential strain before conversion.

θ_i is the critical plane angle. ($0° \leq \theta_i \leq 180°$),

σ'_{11}, ε'_{11}, and γ'_{12} are coordinate normal stress, normal strain, and shearing strain together with θ_i.

By the way of numerical trial method, put stress and strain before conversion into Eqs. (2.23), (2.24), and (2.25), and change continuously the value of θ_i, the normal stress and normal strain corresponding to the new plane will be worked out. The initial value of θ_i is 0° and the upper bound is 180°. Under the theory of

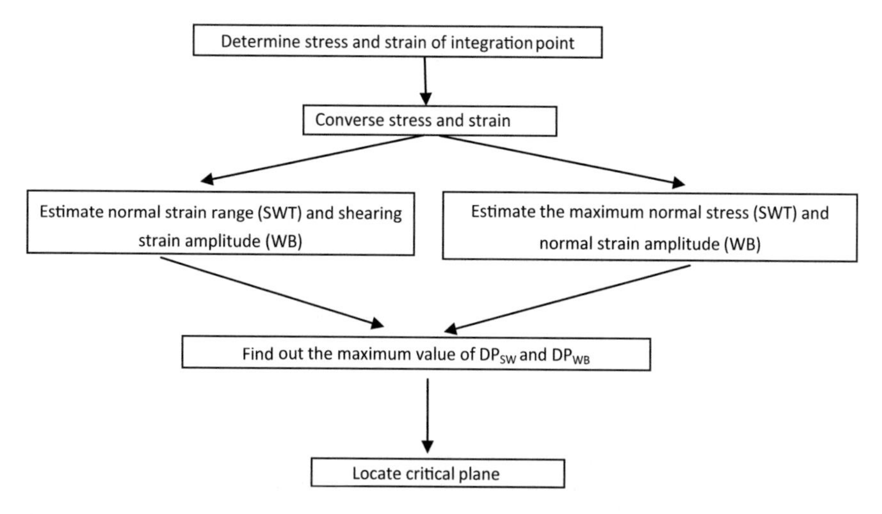

Fig. 2.7 Flow of determining critical plane and damage parameter in SWT and WB model

Smith–Watson–Topper, the angle θ_i corresponding to the maximum normal stress is the critical plane angle and variation range of normal strain on the critical plane is calculated.

The way of locating critical plane and determining damage parameter based on Smith–Watson–Topper (SWT) and Wang–Brown (WB) is shown in flow chart (Fig. 2.7). Wang–Brown model is helpful in finding out the maximum shearing strain and the related critical plane angle θ_i, and calculating the variation range of shearing strain and normal strain on the critical plane.

References

1. Wang, Min. 1999. *Principle and Technology of Anti-fatigue Manufacture*. Jiangsu: Jiangsu Science and Technology Press.
2. Ya, Weixing. 2003. *Analysis of Structure Fatigue Life*. Beijing: National Defense Industry Press.
3. Chen, Chuanrao. 2002. *Fatigue and Fracture*. Wuhan: Huazhong University of Science and Technology Press.
4. Xiong, Junjiang. 2008. *Engineering of Fatigue Fracture Reliability*. Beijing: National Defense Industry Press.
5. Liu, Xinling, Zheng Zhang, and Chunhu Tao. 2010. *Quantitative Analysis of Fatigue Fracture*. Beijing: National Defense Industry Press.
6. Liu, Jun, Wu Henggui, Jinjie Yang, and Zhufeng Yue. 2013. Effect of edge distance ratio on residual stresses induced by cold expansion and fatigue life of TC4 plates. *Engineering Fracture Mechanics* 109: 130–137.

Chapter 3
Effect of Surface Quality of Open Holes on Fatigue Life

Much of the structural loading in an aircraft is transferred through fuselage skin panels to riveted fastening holes at lap joint of these panels. Stress concentration caused by these fastening holes lead to a large loading of cycle tensile stress on these holes, and then bring in fatigue fracture [1]. According to statistics, fatigue fracture of fastening holes accounts for 90% of the whole fatigue fracture of the aircraft, which becomes the main cause of aircraft structure failure. It is a significant research on how to reduce effects of stress concentration to improve aircraft anti-fatigue in design material and manufacture. In this chapter, how to determine the effects of hole surface quality to the stress distribution and life of aluminum perforated panel by finite element software and test are introduced.

3.1 Effect of the Surface Defects on Fatigue Life of Open Holes

Aluminum is the most commonly used material in aircraft structure. It is with high strength and high toughness. But when drilling, it easily introduces surface defects around holes that will increase fatigue crack initiation and crack growth rate on service. Domestic and international scholars and technologists have paid attention to these surface defects and carry out researches on them [2–6].

However, these researches are restricted to improve processing craft to improve surface quality. They have some qualitative analysis on effect of surface defects to fatigue life, without further research.

In this section, effects of three common surface defects to the stress field of center-perforated 2A12-T4 aluminum alloy panel are analyzed, with the help of commercial finite element software ABAQUS. They are scratch, cavity, and inclusion. Also, the model fatigue life is estimated by the way of nominal stress method.

J. Liu et al., *Long-Life Design and Test Technology of Typical Aircraft Structures*,
https://doi.org/10.1007/978-981-10-8399-0_3

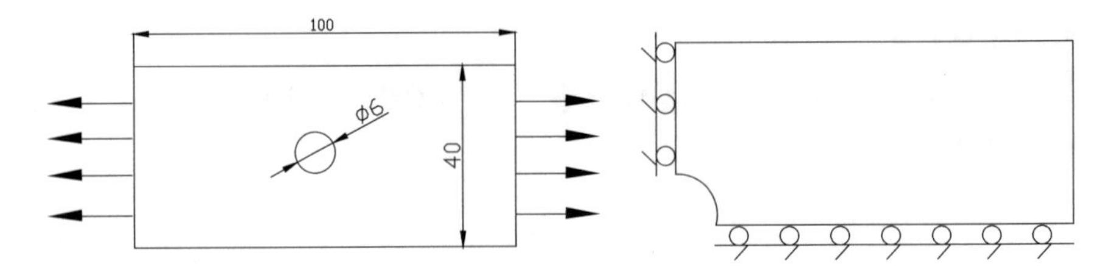

Fig. 3.1 Model of panel with an open hole

The finite element model was a plate with a center hole. The geometries of the model were shown in Fig. 3.1. Uniaxial tension loading on each side of the model was 100 MPa. Due to geometry and loading symmetry with two mid-planes, only one-quarter of the plate was needed and the planes of symmetry bore the related constraint. The test determined elasticity modulus and Poisson ratio of 2A12-T4 aluminum. E = 68 Gpa, μ = 0.3.

3.1.1 Scratch

When drilling holes, scratches would be generated on holes due to chatter of bit or tool backlash movement or other reasons. Scratches have various depth, profile, and directions in reality. In this chapter, in order to study the effect of scratch on stress distribution and fatigue life of hole, scratch was simplified as a column one that penetrates throughout thickness direction. We assumed that the center axis of column is located at profile of hole, which means that the depth of scratch is equal to the radius of column.

In the ABAQUS element library, hexahedron reduced integration elements, C3D8R (three-dimensional eight-nodded continuum elements) for finite element estimate of longitudinal scratch (shown in Fig. 3.2). In order to further study the effect of longitudinal scratch to stress field, we detailed gradually the unit nearby for four times as shown in Fig. 3.3. After detailing, we got 4428 units and 5712 nodes. We calculated 15 models for analysis. There was one model without longitudinal scratch, seven models with 50 μm depth longitudinal scratches, and seven models with 100 μm depth longitudinal scratches. Seven typical longitudinal scratches were chosen that made separately an angle of 0°, 30°, 45°, 60°, 70°, 80° or 90° with horizontal axis (shown in Fig. 3.4). Locations and depths of scratches in model are listed in Table 3.1.

After FEM analysis, the local stress distributions of holes in the 14 models with longitudinal scratch were shown in Fig. 3.5.

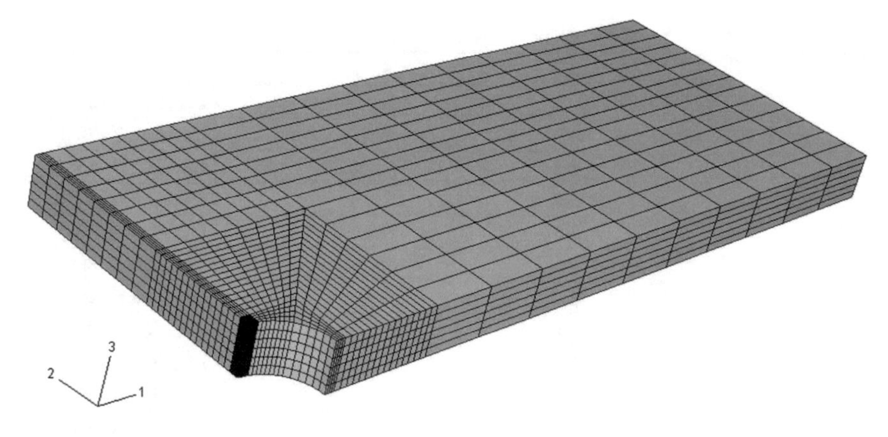

Fig. 3.2 Finite element model

Fig. 3.3 Refiner mesh surround scratch (90°)

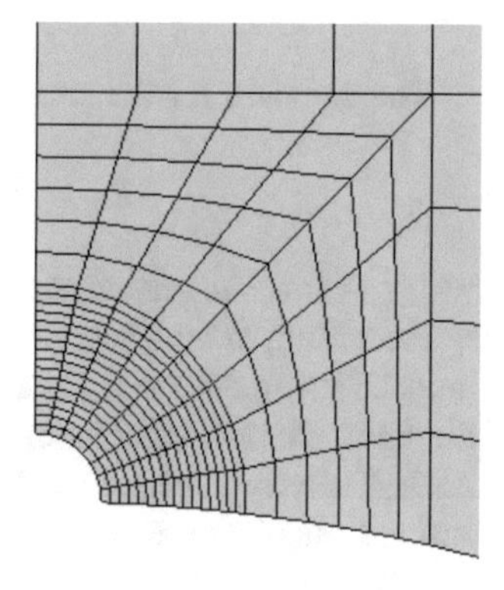

Fig. 3.4 Location of scratch

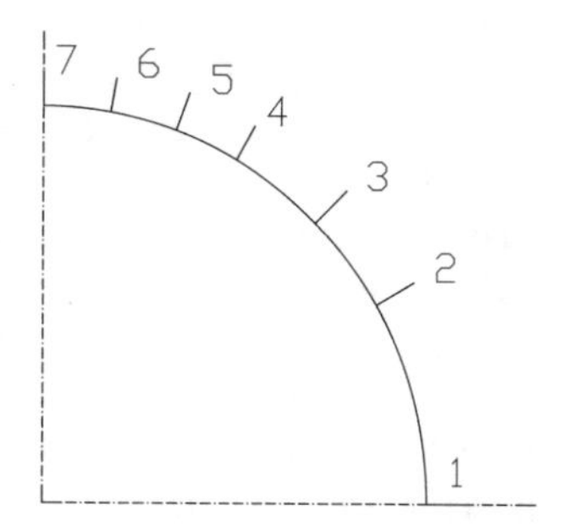

Table 3.1 Location and depth of scratch

Name of model	Depth of scratch	Location of scratch
A0	No scratch	
A1	50 μm	Location 1 (0°)
A2	50 μm	Location 2 (30°)
A3	50 μm	Location 3 (45°)
A4	50 μm	Location 4 (60°)
A5	50 μm	Location 5 (70°)
A6	50 μm	Location 6 (80°)
A7	50 μm	Location 7 (90°)
B1	100 μm	Location 1 (0°)
B2	100 μm	Location 2 (30°)
B3	100 μm	Location 3 (45°)
B4	100 μm	Location 4 (60°)
B5	100 μm	Location 5 (70°)
B6	100 μm	Location 6 (80°)
B7	100 μm	Location 7 (90°)

The equation for the stress concentration factor of the center plate is

$$K_{tg} = 2 + 0.15(D/W)^2 + (1 - D/W)^3, \tag{3.1}$$

where D is the diameter of the hole and W is the width of the plate.

According to Eq. (3.1), we can get the theoretical concentration coefficient of the model, $D/W = 6/40 = 0.15$, $K_{tg} = 2.6175$. This is basically the same as the result of the finite element calculation of model A0 (see Fig. 3.5a) with a relative error of 2.3%. Therefore, the grid and the boundary conditions of the model are reasonable, and the calculation results are credible.

From stress distributions of these holes with scratches, it can be seen that when depth of scratch is 50 μm and its location is location 1, 2, 3 or 4, the point of maximum stress and its value are separately same with or similar to those of the model with no scratch. It means that at these locations, scratches have no clear effect on stress distribution of hole. When scratch is at location 5, 6, or 7, the maximum stress occurs at the root of scratch and the values are 282, 312 and 325 MPa, respectively. It shows that the scratches have a great effect on the maximum stress. So scratches with depth of 50 μm have effect on the location and value of the stress distribution of hole when it is in area of 70°–90°.

When depth of scratch is 100 μm and its location is location 1, 2, or 3, the point of maximum stress and its value are separately same with or similar to those of the model with no scratch. It means that at these locations, scratches have no clear effect on stress distribution of hole. When location of scratch is location 4, 5, 6 or 7, the maximum stress occurs at the root of scratch and their values are 276, 312, 325, and 341 MPa. So scratches with depth of 100 μm have effect on the stress distribution of hole only when it is in area of 60°–90°. In general, fatigue damage occurs

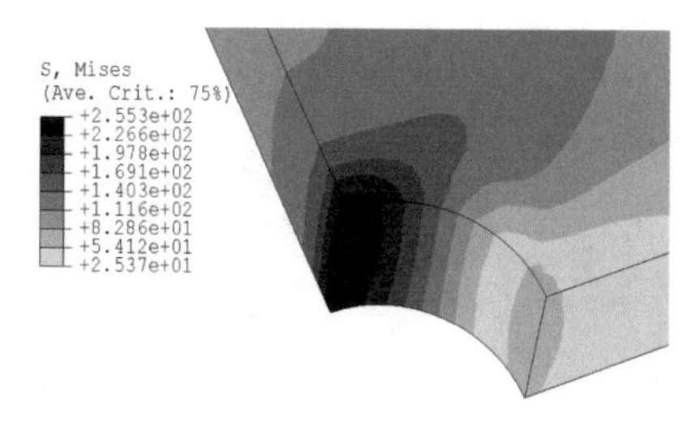

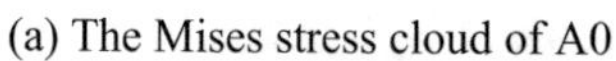
(a) The Mises stress cloud of A0

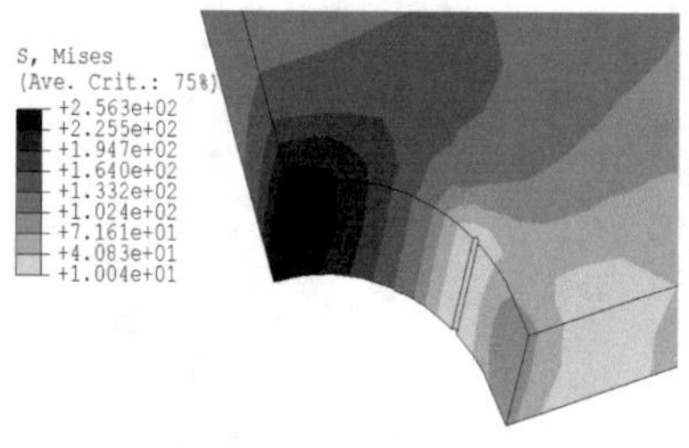

(b) The Mises stress cloud of A1

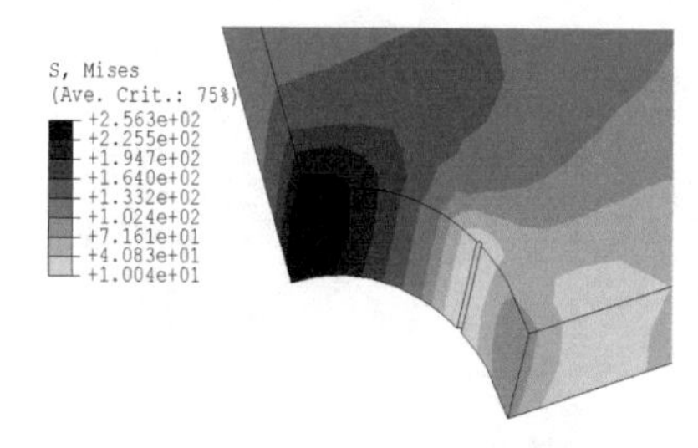

(c) The Mises stress cloud of A2

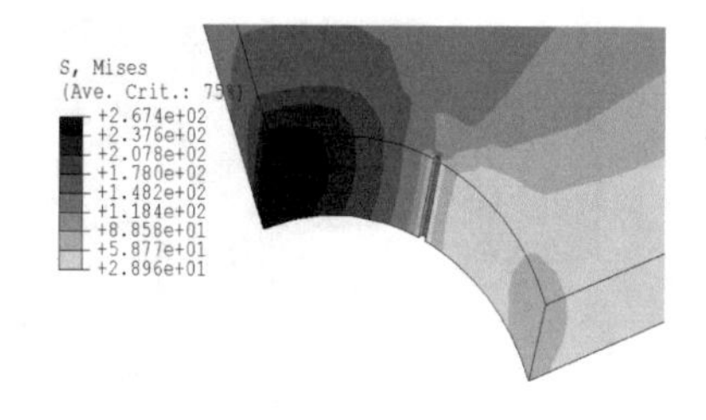

(d) The Mises stress cloud of A3

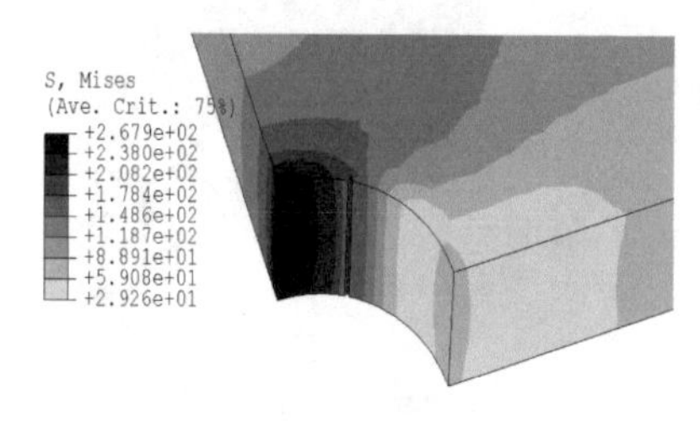

(e) The Mises stress cloud of A4

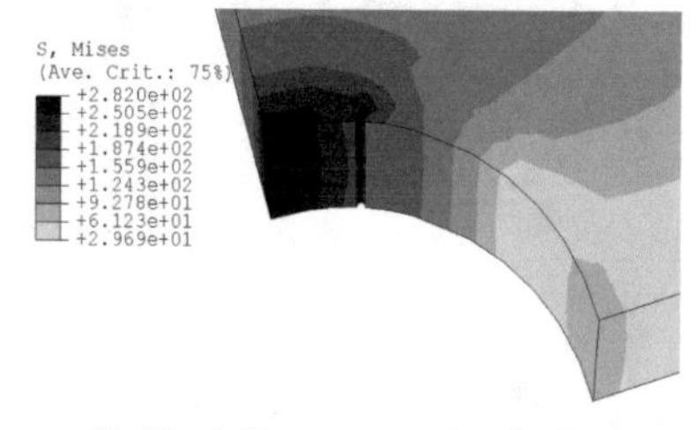

(f) The Mises stress cloud of A5

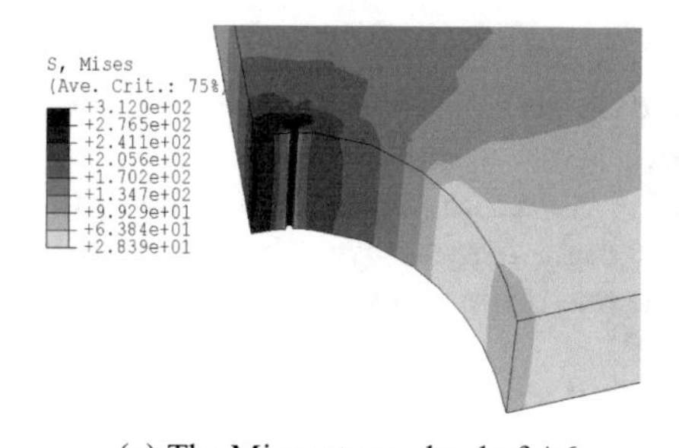

(g) The Mises stress cloud of A6

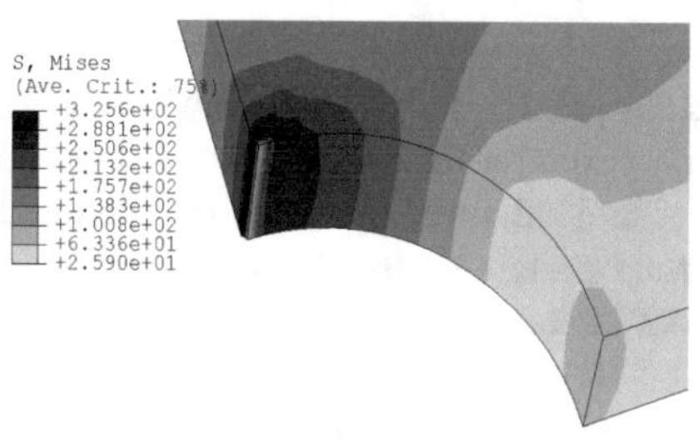

(h) The Mises stress cloud of A7

Fig. 3.5 The fracture hole stress cloud

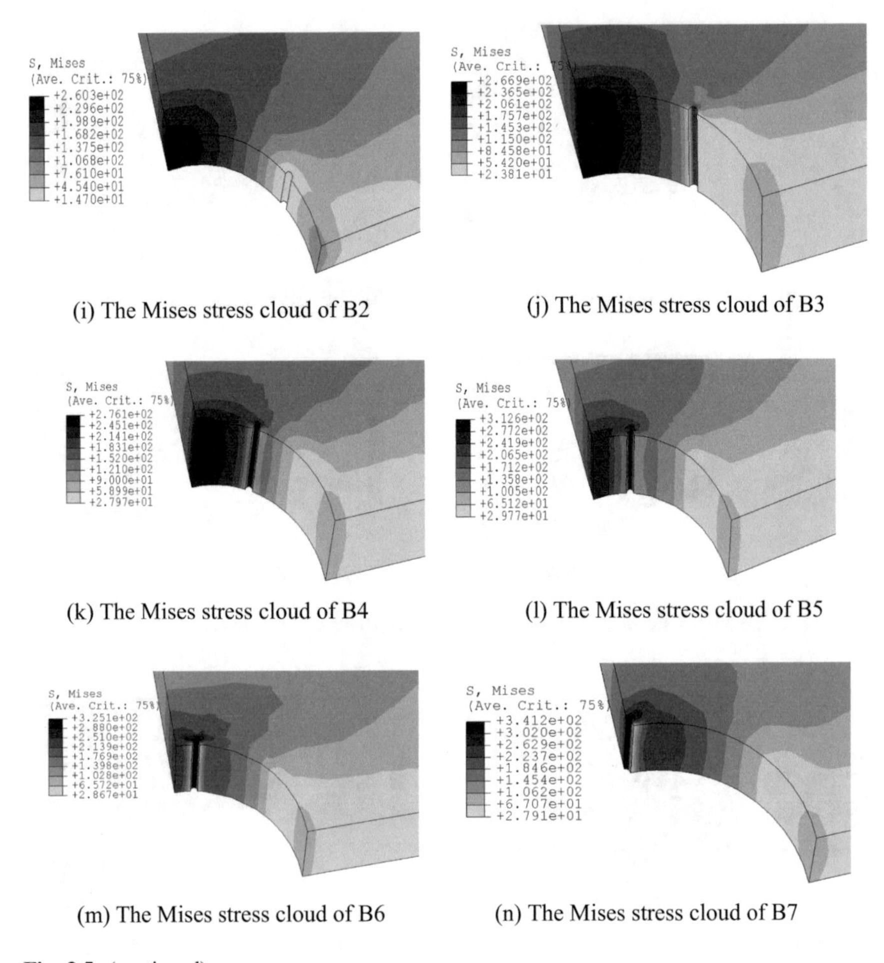

(i) The Mises stress cloud of B2

(j) The Mises stress cloud of B3

(k) The Mises stress cloud of B4

(l) The Mises stress cloud of B5

(m) The Mises stress cloud of B6

(n) The Mises stress cloud of B7

Fig. 3.5 (continued)

at the point with the maximum stress. Based on results of the above models, stress concentration factor K_T can be determined.

Nominal stress approach intruded in 1.4.1 is used to predict the fatigue life of center-perforated panel with scratches. Assuming that the cycle load is constant amplitude load, $R = S_{\min}/S_{\max} = -1$ and the maximum nominal stress is 100 MPa, we will get $S_a = (S_{\max} - S_{\min})/2 = 100$ MPa.

S-N curves of 2A12T4 aluminum alloy with $R = -1$ and various K_T are shown in Fig. 3.6 [7].

If they are constant amplitude loads ($R = 1$), fatigue life of models can be predicted by polynomial interpolating function, according to *S-N* curves, as shown in Fig. 3.6. And then the relationship curve between location of scratch and logarithmic fatigue life can be drawn (see Fig. 3.7).

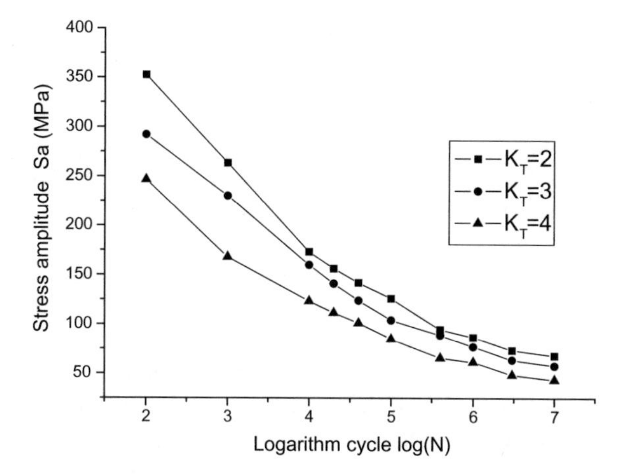

Fig. 3.6 *S-N* curve of 2A12T4 aluminum alloy

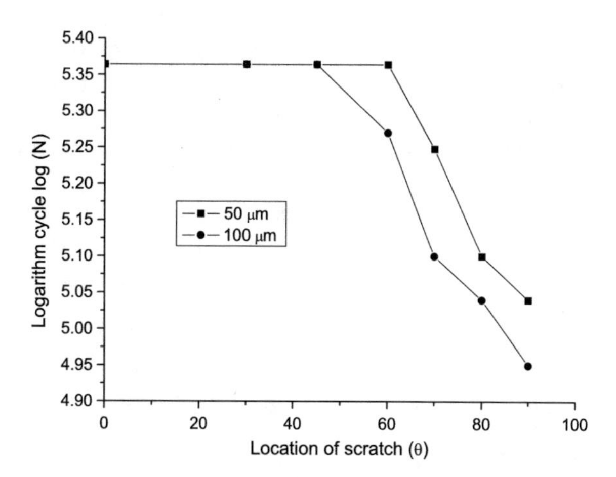

Fig. 3.7 Fatigue life versus location of scratch

Figure 3.7 shows that when the depth of scratches is 50 μm, only if they are in the area of 70°–90°, stress life of the fastening holes will decrease with the increase of the angle. If the location is in the area of 90°, fatigue life is the shortest, nearly 47% of original life. When the depth of scratches is 100 μm, only if they are in the area of 60°–90°, stress life of the fastening holes will decrease with the increase of the angle. If the location is in the area of 90°, fatigue life is the shortest, nearly 39% of original life.

Therefore, there exists an effected area centering on the maximum stress of the hole without scratch. Only if scratches are in this area, fatigue life of fastening hole will be affected by them. With the growth of depth, the area expands outside. In this area, the more the depth is, the shorter the fatigue life is. The closer to center the scratch is, the shorter the fatigue life is.

3.1.2 Cavity

Because of the chemical reaction of the plate in casting or other reasons, there will be cavity on the surface of the alloy. If the alloy plate with cavities on surface bears high load, the cavities are equivalent to notch. Around the cavities, stress concentration will occur, forming fatigue crack source, so that fatigue life of the component is reduced.

The problem of stress distribution on the surface of holes with cavities can be regarded as the plane stress problem. The real cavities have different shapes. Here, in order to study the effect of cavity on the stress distribution and fatigue life of the hole, the cavity is simplified to tiny circles with its center on the hole's surface. Thus, the radius of the cavity is the radius of the tiny circle.

In the finite element calculation, integral unit (CPS4R) reduction by four-node plane stress is used in the ABAQUS cell library. The grid is shown in Fig. 3.8. In order to study the effect of cavity on the stress field, we detailed gradually the unit nearby for four times as shown in Fig. 3.9. We calculated 15 models for analysis. There was one model without cavities, seven models with 50 μm depth cavities, and seven models with 100 μm depth cavities. Seven typical cavities were chosen that made separately an angle of 0°, 30°, 45°, 60°, 70°, 80° or 90° with horizontal axis (shown in Fig. 3.4). Locations and depths of cavities in model are shown in Table 3.2.

The symmetrical constraint U1 = UR2 = 0 in direction 1 is applied to the left end of the model. The symmetrical constraint U2 = UR1 = 0 in direction 2 is applied to bottom of the model, and the symmetric tension of the right end is 100 N/mm.

After the finite element analysis, the local Mises stress cloud of the orifices of models with cavities is shown in Fig. 3.10.

The stress distribution of the orifice shows:

When the cavity radius is 50 μm and it is at position 1, 2, 3 or 4, the point of maximum stress and its value are separately same with or similar to those of the model with no inclusion. It can be assumed that cavities have no effect on the maximum stress distribution of hole. When cavity is at location 5, 6, or 7, the

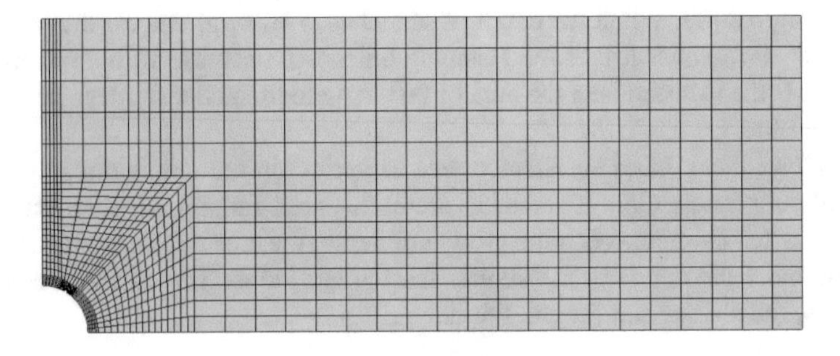

Fig. 3.8 Finite element model of the central perforated plate with cavities

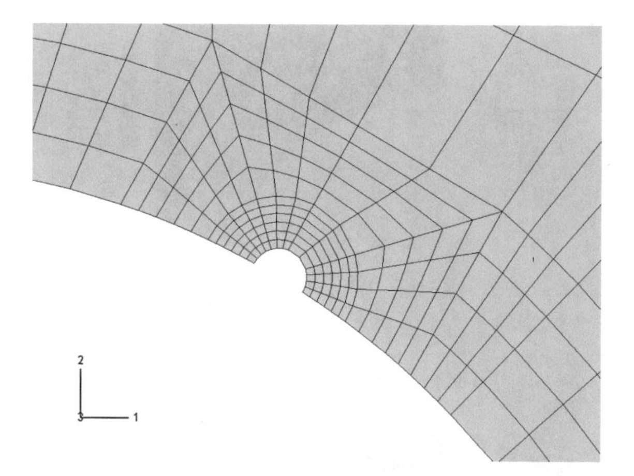

Fig. 3.9 Refined grids near cavities

Table 3.2 Cavity size and location

Model code	Cavity radius (μm)	Cavity position
C1	50	Location 1 (0°)
C2	50	Location 2 (30°)
C3	50	Location 3 (45°)
C4	50	Location 4 (60°)
C5	50	Location 5 (70°)
C6	50	Location 6 (80°)
C7	50	Location 7 (90°)
D1	100	Location 1 (0°)
D2	100	Location 2 (30°)
D3	100	Location 3 (45°)
D4	100	Location 4 (60°)
D5	100	Location 5 (70°)
D6	100	Location 6 (80°)
D7	100	Location 7 (90°)

maximum stress occurs at the root of the cavity and the values are 281, 312, and 334 MPa, respectively. It shows that cavities have a great effect on the maximum stress. So cavities have effect on the location and value of the stress distribution of hole when it is in area of 70°–90°.

When depth of cavities is 100 μm and its position is location 1, 2, or 3, the point of maximum stress and its value are separately same with or similar to those of the model with no cavities. It can be assumed that cavities have no effect on the maximum stress distribution and its value. If cavity is at location 4, 5, 6 or 7, the maximum stress occurs at the root of cavity and their values are 281, 312, 325 and 342 MPa, respectively. So cavities have effect on the stress distribution and its value only if it is in area of 60°–90°.

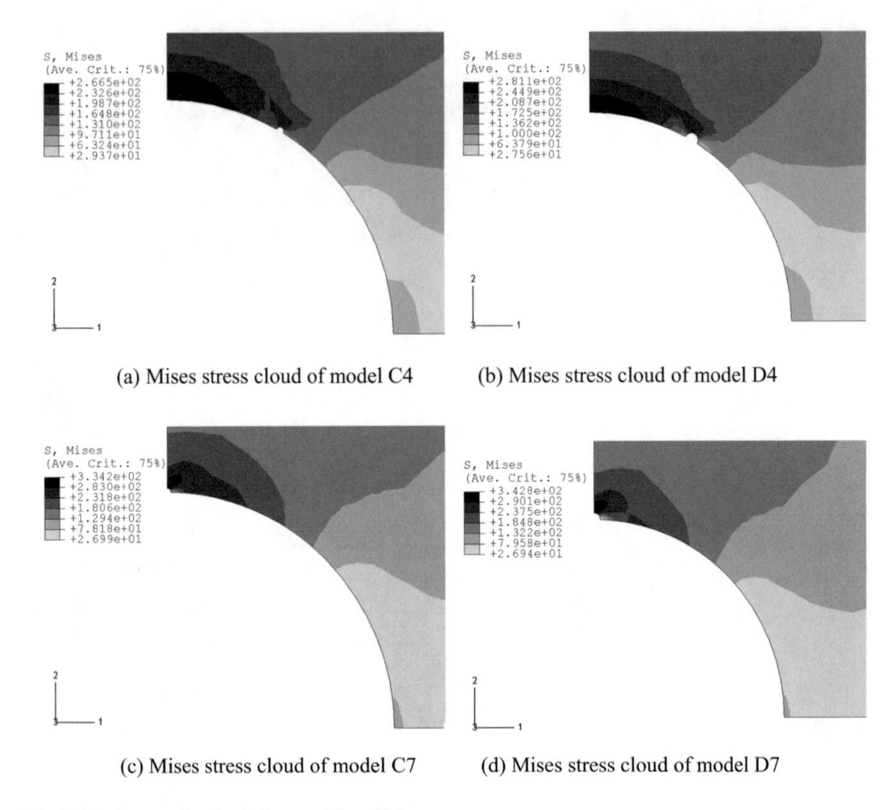

Fig. 3.10 Stress cloud of plane with cavities

Based on results of the above models, nominal stress approach is used to predict the fatigue life of center-perforated panel with cavities. The results are shown in Table 3.3.

It can be found that when the cavity radius is 50 μm, only if the cavity is in the area of 70°–90° the fatigue life of the fastening hole decreases along with the increase of the angle. When cavity is in the area of 90°, the fatigue life is shortest nearly 42% of the original life. When the cavity radius is 100 μm, only if the cavity is in the area of 60°–90°, the fatigue life of the fastening hole decreases with the increase of the angle. When the cavity is in the area of 90°, the fatigue life is the shortest nearly 38% of the original life.

It can be seen that there is also an affected area centering on the maximum stress point of fastening hole without cavities. The cavity will have an effect on the fatigue life of the fastening hole only if it is in this area. With the increase of the cavity size, the affected area will expand outward. In this affected area, the larger the cavity is, the shorter the fatigue life of the hole is. The closer to the center of the affected area the cavity is, the shorter the fatigue life of the hole is.

Table 3.3 Fatigue life of holes with cavities

Model code	Nominal stress σ/ MPa	Stress concentration factor K_T	Fatigue life N/ cycle
C1	100	2.55	231,206
C2	100	2.55	231,206
C3	100	2.55	231,206
C4	100	2.55	231,206
C5	100	2.82	176,929
C6	100	3.12	125,892
C7	100	3.34	98,138
D1	100	2.55	231,206
D2	100	2.55	231,206
D3	100	2.55	231,206
D4	100	2.81	178,816
D5	100	3.12	125,892
D6	100	3.25	109,647
D7	100	3.42	88,890

3.1.3 Inclusion

There are more or less some impurities in materials in its metallurgical process, such as alumina, magnesium oxide, and so on. When the two-phase particles are well bonded with the body and the matrix phase is strong, it has little effect on the crack propagation. However, when the two-phase particles are relatively large and are weakly bonded to the substrate, and there is stress concentration around them, voids are likely to be formed around them, resulting in acceleration of crack growth and reduction of fatigue life. Therefore, these impurities in the surface of the component will also have a certain impact on the surface cracks, reducing the fatigue life of components. The location, size, quantity, and geometry of the inclusion have a significant effect on properties of material [8, 9].

The problem of stress distribution on the surface with inclusions can also be regarded as the plane stress problem. The real inclusions have different shapes. Here, in order to study the effect of inclusions on the stress distribution and fatigue life of the hole, inclusion is simplified as a circular circle with its center on the contour of the hole, just like cavity is. Thus, radius of the circular inclusions is the radius of the tiny circle. The inclusion compound Al_2O_3 is treated as elastic material with elastic modulus $E = 3.9 \times 10^5$ MPa and Poisson's ratio $\mu = 0.25$ [10].

In the finite element calculation, integral unit (CPS4R) reduction by four-node plane stress is used in the ABAQUS cell library. The grid is shown in Fig. 3.8. In order to study the effect of inclusion on the stress field, we detailed gradually the unit nearby for four times as shown in Fig. 3.11. We calculated 15 models for analysis. There was one model without inclusion, seven models with 50 μm radius inclusion, and seven models with 100 μm radius inclusion. Seven typical inclusions

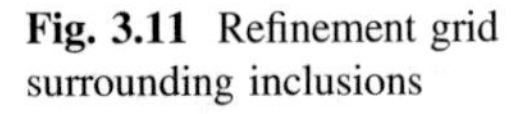

Fig. 3.11 Refinement grid surrounding inclusions

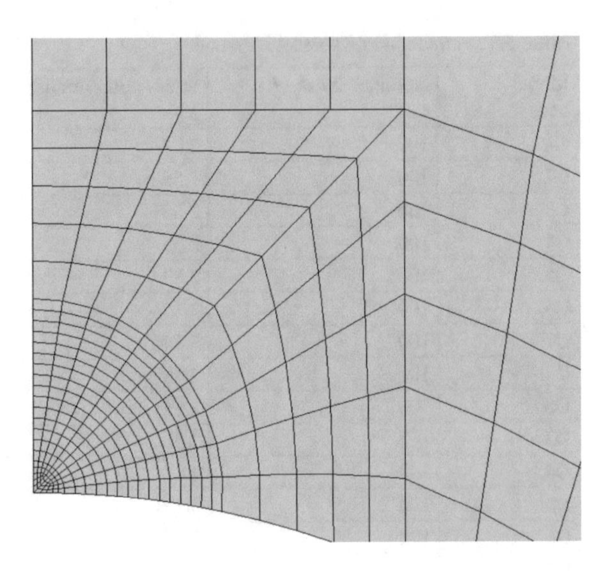

Table 3.4 Inclusion size and location

Model code	Hole radius (μm)	Hole position
E1	50	Location 1 (0°)
E2	50	Location 2 (30°)
E3	50	Location 3 (45°)
E4	50	Location 4 (60°)
E5	50	Location 5 (70°)
E6	50	Location 6 (80°)
E7	50	Location 7 (90°)
F1	100	Location 1 (0°)
F2	100	Location 2 (30°)
F3	100	Location 3 (45°)
F4	100	Location 4 (60°)
F5	100	Location 5 (70°)
F6	100	Location 6 (80°)
F7	100	Location 7 (90°)

were chosen that made separately an angle of 0°, 30°, 45°, 60°, 70°, 80° or 90° with horizontal axis (shown in Fig. 3.4). Locations and depths of inclusions in model are shown in Table 3.4.

The symmetrical constraint U1 = UR2 = 0 in direction 1 is applied to the left end of the model. The symmetrical constraint U2 = UR1 = 0 in direction 2 is applied to the bottom of the model. The symmetrical tension on right end of the model is 100 N/mm.

After finite element analysis, the local Mises stress cloud image of the open-hole model with inclusions is shown as in Fig. 3.12 (without inclusions).

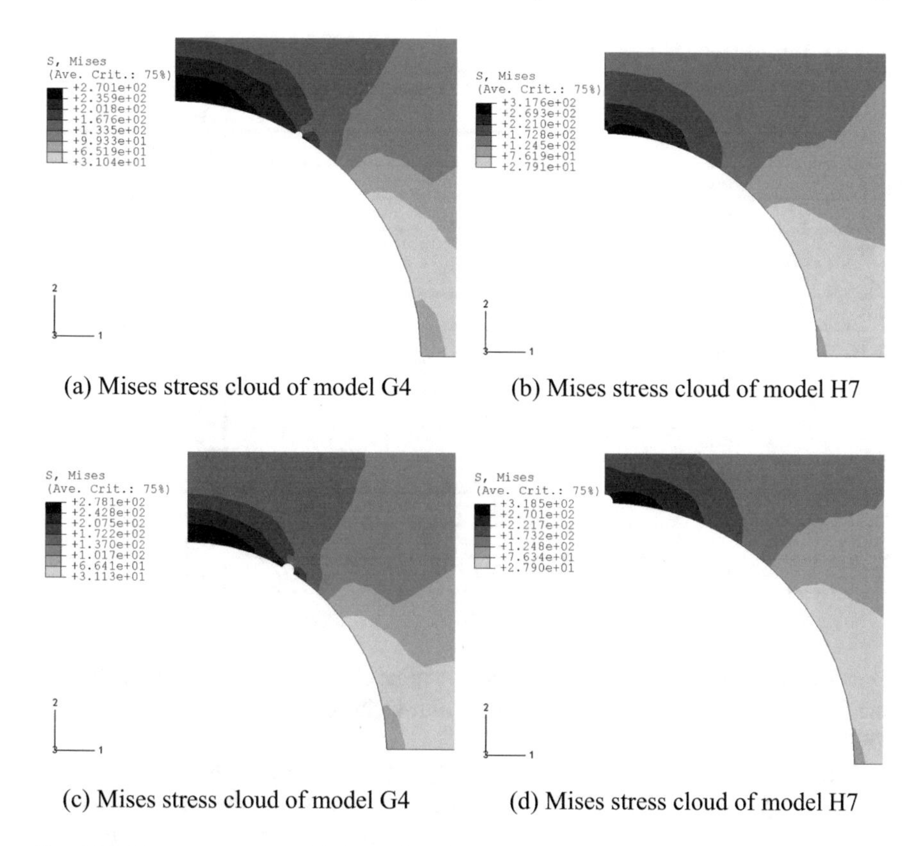

(a) Mises stress cloud of model G4 (b) Mises stress cloud of model H7

(c) Mises stress cloud of model G4 (d) Mises stress cloud of model H7

Fig. 3.12 Stress distribution of holes with inclusions

Stress distribution of holes with inclusions shows:

When the inclusion radius is 50 μm and the inclusion is at position 1, 2, 3, or 4, the point of maximum stress and its value are separately same with or similar to those of the model with no inclusion. It can be assumed that inclusions have no effect on the maximum stress distribution of the hole. When inclusion is at location 5, 6 or 7, the maximum stress occurs at the root of the inclusion and the values are 278, 304, and 317 MPa, respectively. It shows that inclusions have a great effect on the maximum stress. So inclusions have effect on the location and value of the stress distribution of hole when it is in area of 70°–90°.

When inclusion depth is 100 μm and the cavity is at location 1, 2, or 3, the point of maximum stress and its value are separately same with or similar to those of the model with no inclusions. It can be assumed that inclusions have no effect on the maximum stress distribution and its value. If inclusion is at location 4, 5, 6 or 7, the maximum stress occurs at the root of inclusion and their values are 278, 296, 313 and 319 MPa. So inclusions have effect on the stress distribution and its value only if it is in area of 60°–90°.

Table 3.5 Fatigue life of holes with inclusions

Model code	Nominal stress σ/MPa	Stress concentration factor K_T	Fatigue life N/cycle
E1	100	2.55	231,206
E2	100	2.55	231,206
E3	100	2.55	231,206
E4	100	2.55	231,206
E5	100	2.78	184,459
E6	100	3.04	139,474
E7	100	3.17	120,225
F1	100	2.55	231,206
F2	100	2.55	231,206
F3	100	2.55	231,206
F4	100	2.78	184,459
F5	100	2.96	152,380
F6	100	3.13	125,925
F7	100	3.19	117,449

Based on results of the above models, nominal stress approach is used to predict the fatigue life of center-perforated panel with inclusions. The results are shown in Table 3.5.

When the inclusion radius is 50 μm, only if the inclusion is in the area of 70°–90° the fatigue life of the fastening hole decreases along with the angle increasing. When inclusion is in the area of 90°, the fatigue life is shortest nearly 52% of the original life. When the inclusion radius is 100 μm, only if the inclusion is in the area of 60°–90°, the fatigue life of the fastening hole decreases with the increase of the angle. When the inclusion is in the area of 90°, the fatigue life is the shortest nearly 51% of the original life.

It can be seen that there is also an affected area centering on the maximum stress point of fastening hole without inclusion. The inclusion will have an effect on the fatigue life of the fastening hole only if it is in this area. With the increase of the inclusion size, the affected area will expand outward. In this affected area, the larger the inclusion is, the shorter the fatigue life of the hole is. The closer to the center of the affected area the inclusion is, the shorter the fatigue life of the hole is.

In addition, Fig. 3.13 shows the fatigue life changes of holes affected by inclusion and cavity in the same size (holes R = 50 μm and R = 100 μm, respectively). It can be seen that the influence of the cavity on the fatigue life of the hole is slightly larger than that of the inclusion in the same size.

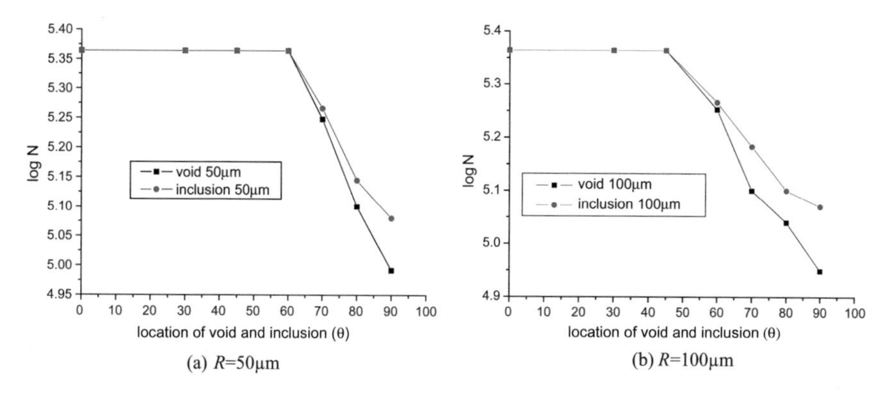

Fig. 3.13 Cavity and inclusion position and fatigue life curve of pore

3.2 Effect of Manufacturing Quality on Fatigue Performance of Open Holes

In the case of drilling, although the borehole is a basic material removal process, the mechanical action of the borehole causes plastic deformation in the area near the surface of the hole and the drill bit marks and scales on the surface of the hole when it leaves, in which stress concentration can be generated. Even small surface defects can increase stress and increase the chances of crack generation and accelerate the rate of fatigue crack propagation. These surface defects have attracted people's attention, and they have carried out some research [11–14], Pei Xuming, from the perspectives of surface integrity and fatigue life, conducted an experimental and numerical Simulation study on fastening hole surface quality of aircraft made of 7075 aluminum alloy. It shows that the appropriate reduction in feed rate and increase of cutting speed can improve the surface quality of the fastening holes.

However, these researches are limited to how to improve the processing technology to improve the surface quality of the orifice. They have some qualitative analysis on effect of surface quality in manufacturing processing to fatigue life, without further research.

Carter studied the effects of different drilling processes on the machining quality of the holes and gave four parameters describing the quality of the holes: surface roughness, cylindricality, quantity, and direction of furrows [3]. However, these parameters do not fully define the machining quality of the hole, such as roundness, a machining quality parameter for the hole.

Most of the processing quality parameters are a statistic. Two test pieces with the same processing quality parameters have different surface contour or shape. And the ideal shape of the specimen is difficult to obtain. In addition, if the specimen has been processed, these parameters are present simultaneously. It is difficult to quantitatively analyze the impact of a parameter on the fatigue life of the hole. Therefore, it is very difficult to study the effect of different processing quality parameters on structural fatigue performance by using the experimental method, and it must be solved together with other means.

In this section, we study effects of four commonly used processing quality parameters describing holes on the fatigue life of openings. They are roughness, verticality, cylindricality, and roundness.

3.2.1 Surface Roughness

Surface roughness is an important parameter for assessing surface integrity. The formation of surface roughness is mainly due to the friction between tool and parts, plastic deformation and metal tear in cutting separation, and the existence of high-frequency vibration and other reasons. Machining methods, machine tool precision, vibration and adjustment conditions, the geometric parameters of the cutting tool and its wear, cutting, cooling, and lubrication will have a direct impact on the roughness of the machined surface [13].

The effects of roughness on fatigue life were studied by means of the experimental method. The fatigue specimens with different roughness were the central perforated aluminum plate with thickness of 4 mm and the length direction was the rolling direction. The specific dimensions are shown in Fig. 3.14. According to the specimen size, combined with Eq. (3.1), we can see that the theoretical stress concentration coefficient $K_T = 2.52$.

Taking into account the dispersion of the data of the fatigue experiment, a total of 21 specimens were tested. Although the drilling looks like a simple process, in fact, it is a more complex process. The drill bit is a complex spiral of geometries and is deformable in the axial direction. Compared with other processes, such as turning, milling, grinding, working plane of the drill bit is inside the workpiece, and the scraps must be removed from the borehole, otherwise, it will hinder the drilling lubrication and cooling. Other processing qualities (such as cylindrical degrees) of the holes may be different for two holes with the same roughness due to the condition of drill bit [3]. In order to eliminate the influence of these factors on the fatigue life of the hole, the same machine and new drill is used in processing, and the speed of the tool is kept constant and completed by the same worker. At this

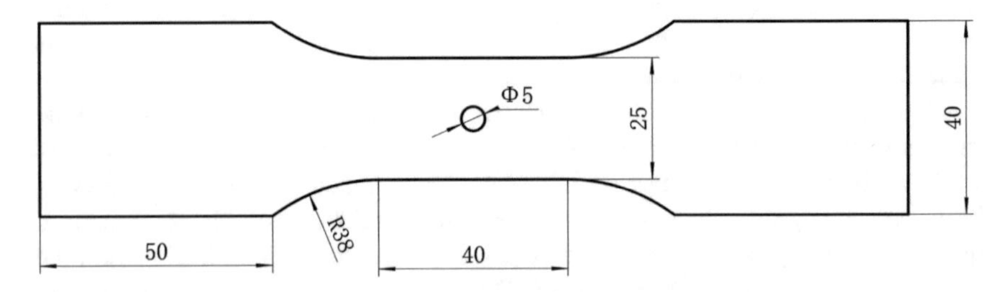

Fig. 3.14 Specimen size

Table 3.6 Specimen holes-processing parameters

	Group A	Group B	Group C
Spindle speed	480 r/min		
Tool diameter	5 mm		
Feed rate	24 mm/min	12 mm/min	6 mm/min

time, the roughness of the hole can be determined by the feed rate of the tool [14]. Because aluminum alloy is softer compared with drill bit, in the processing of the 21 test pieces, the drill wear is ignored. Therefore, it can be considered that these 21 specimens form a different hole surface roughness only by the different feed speed of the tool, with the other processing quality basically unchanged.

According to different feed speed, the 21 test pieces are divided into A, B, and C three groups, and each group contains seven. The tool feed rate of group A is 24 mm/min, group B is 12 mm/min, and group C is 6 mm/min. The detailed process parameters are shown in Table 3.6.

The current assessment of surface roughness is qualitative and quantitative, qualitative assessment determines the surface roughness of the measured parts by means of block or magnifying glass and microscope, and cannot draw a specific contour micro-roughness of the value; quantitative assessment can accurately determine the value of specific micro-roughness of contour by means of modern detection equipment.

The test of the comparison method is to judge the surface roughness parameters by comparing surface of the specimen with working surface of roughness block by vision and touching. The material, shape and manufacturing process of the roughness sample should be as close as possible to the workpiece to be inspected. As long as the comparison sample qualified and the examiner is experienced, we will get satisfactory results by the way of comparison to test the surface roughness of the workpiece. Therefore, this method is commonly used in the production site.

In order to facilitate the engineering application, the method of determining the surface roughness of the hole is a comparison method commonly used in the factory. After measurement, the roughness of group A was $Rz = 6.3$, the roughness of group B was $Rz = 3.2$, and the roughness of group C was $Rz = 1.6$.

The roughness fatigue test was performed on the INSTRON8802 electrohydraulic servo testing machine. Load mode was stress control, with sine fatigue load spectrum of wave amplitude, stress ratio 0.1, loading frequency 16 Hz and the maximum nominal stress 120 MPa. The test was carried out at room temperature. The connection between the open specimen and the test machine is shown in Fig. 3.15.

After the test, the median fatigue life, standard deviation, and coefficient of variation of each group are shown in Table 3.7, where the coefficient of variation reflects the dispersion of the data.

It can be seen that the smaller the surface roughness of the opening is, the longer its fatigue life is. Compared with life when $Rz = 6.3$, the lifespan increases separately by 1.5 and 1.9 times when $Rz = 3.2$ and $Rz = 1.6$. Therefore, reducing the surface roughness of the pores can significantly improve the fatigue life of them. And if $Rz = 1.6$, its life dispersibility is minimal.

Fig. 3.15 Connection diagram of the specimen and INSTRON8802 test machine

Table 3.7 Median fatigue life, standard deviation, and coefficient of variation for different roughness specimens

Roughness	Median fatigue life/cycle	Standard deviation	Coefficient of variation
6.3	60,146	7658.55	0.127
3.2	88,651	19424.71	0.219
1.6	115,081	7846.90	0.068

According to the experimental data, the empirical equation of the influence of roughness on the fatigue life of the perforated specimen was established by regression analysis.

$$N = N_0 \cdot e^{-0.136R} \tag{3.2}$$

In Eq. (3.2), N is the fatigue life, N_0 is the fatigue life of the ideal smooth hole, and R is the roughness of the hole.

Scattering electron microscopy was used to analyze the fracture of all fatigue specimens. It is found that as to roughness, there are two types of cracks that cause fatigue failure of the hole: corner cracks and surface cracks. Typical cracks are shown in Fig. 3.16.

The results show that there are three corner cracks in group A and four surface cracks. Group B has two angle cracks and five surface cracks. Group C has 0 corner crack and seven surface cracks. That is, when Rz = 6.3 and Rz = 3.2, the fatigue crack of the hole has two forms, that are corner crack and surface crack, and when Rz = 1.6, the fatigue main crack of the hole exists only in the form of surface crack.

(a) Corner crack (b) Surface crack

Fig. 3.16 Roughness-induced cracking of the open-hole crack

The results are shown in comparison with Table 3.6. In the A and B groups, the fatigue crack of the hole has two forms, corner crack and surface crack, and the corner crack is more easily formed compared with the surface crack. The effect of corner crack on the fatigue life of the perforated specimen is greater than that of the surface crack, resulting in a large dispersion of the data, while in the C group because there is only surface fatigue crack on the hole, its data dispersion is distinctly smaller than those in group A and group B.

3.2.2 *Verticality*

In the production practice, it is found that: after the manual hole, the axial direction of the hole is not exactly the same with the normal direction of the machining surface, but there is a certain angle [3] which is called angle a. In addition, the plane formed by the ideal axis and the actual axis of the hole and the portrait of processing surface form the angle b, as shown in Fig. 3.17. In this chapter, the angles a together with b is regarded as vertical tolerances.

In order to study the effect of verticality on the stress distribution of the hole, a numerical simulation method is adopted. Here, a total of 13 models are analyzed, in which the model D0 is a central perforated plate having an ideal opening. The vertical tolerances for each model are shown in Table 3.8.

The symmetrical constraint U1 = UR2 = 0 in direction 1 is applied to the left end of the model. The symmetrical constraint U2 = UR1 = 0 in direction 2 is applied to the bottom of the model. The symmetrical tension on the right end of the model is 100 N/mm.

After finite element analysis, the distribution of the stress field of each model can be obtained. The stress cloud of the typical model D0 (ideal hole) is shown in Fig. 3.18. According to the stress field distribution of the model, the theoretical

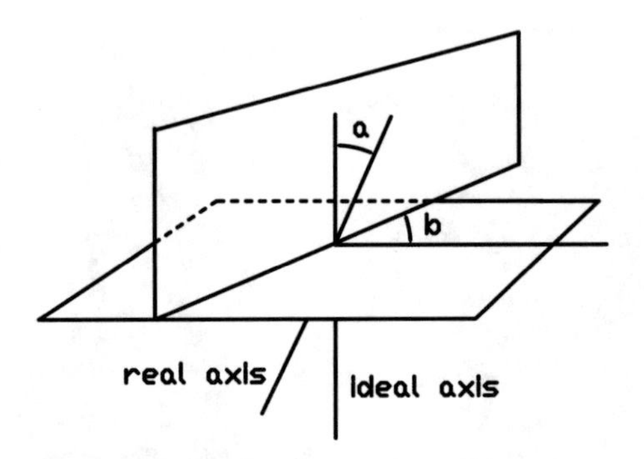

Fig. 3.17 Definition of vertical angle

Table 3.8 Tolerance of the verticality of each model

Model code	Angle a (°)	Angle b (°)
D0		
D1	5°	0°
D2	10°	0°
D3	15°	0°
D4	20°	0°
D5	5°	45°
D6	10°	45°
D7	15°	45°
D8	20°	45°
D9	5°	90°
D10	10°	90°
D11	15°	90°
D12	20°	90°

Fig. 3.18 Stress cloud of model D0 (ideal hole)

Table 3.9 Fatigue life of openings with different vertical degrees

Model code	Nominal stress σ/MPa	Angle a (°)	Angle b (°)	Stress concentration factor K_T	Fatigue life N/cycle
D0	100			2.56	231,663
D1	100	5°	0°	2.56	231,663
D2	100	10°	0°	2.56	231,663
D3	100	15°	0°	2.56	231,663
D4	100	20°	0°	2.56	231,663
D5	100	5°	45°	2.59	222,954
D6	100	10°	45°	2.65	210,283
D7	100	15°	45°	2.7	200,085
D8	100	20°	45°	2.75	190,219
D9	100	5°	90°	2.66	208,217
D10	100	10°	90°	2.72	196,100
D11	100	15°	90°	2.79	182,564
D12	100	20°	90°	2.89	164,349

stress concentration coefficient $K_T = 2.56$ can be obtained. According to Eq. (3.1), the theoretical concentration coefficient of the perforated plate is 2.52. The theoretical solution and the finite element method are basically the same. It can be considered that the loading of the finite element model, the boundary condition and the size of the grid are appropriate, and the calculation result is credible.

Using the nominal stress method, combined with the given *S-N* curves with different K_T, the predicted values of fatigue life of the open-cell with different vertical degrees can be obtained, as shown in Table 3.9.

It can be seen from Table 3.9 that the angle b has the greatest effect on the fatigue life of the hole model. When the angle b is 0°, the fatigue lives of the opening affected by different values of angle a are the same with those of the ideal hole. The angle a has no effect on the fatigue life of the hole. When the angle b is 45°, different values of angle a have distinct effect on fatigue life; when the angle b is 90°, the effect of angle a on the fatigue life of the hole is the greatest. Therefore, angle b determines the fatigue life of the opening. In addition, when angle b is constant and angle a has effects on the opening fatigue life, the fatigue life of the opening decreases gradually along with the angle a increases.

In engineering practice, angle b is related to the loading direction and difficult to obtain in advance. In order to obtain the quantitative relationship between the verticality and the fatigue life of the openings, a conservative assumption is made, that is, angle b is always 90° at work. Based on this assumption, the linear regression method can be used to obtain the empirical equation of the effect of verticality on the fatigue life of the opening:

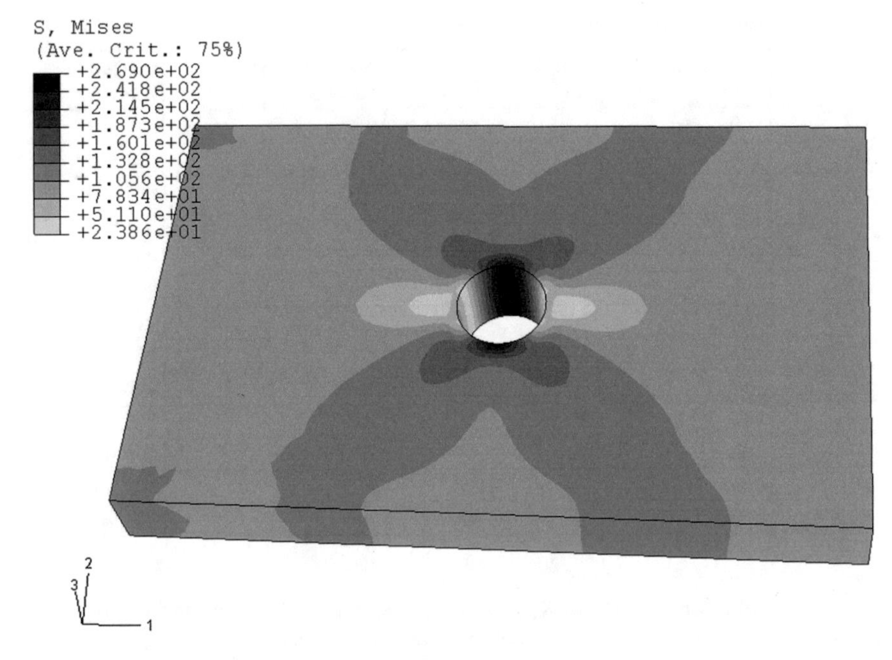

Fig. 3.19 Model stress cloud of model with verticality of $a = 8°$

$$N = N_0 * e^{-0.016a} \tag{3.3}$$

In Eq. (3.3), N is the fatigue life of the hole, N_0 is the fatigue life of the ideal smooth opening and a is the verticality of the hole.

In order to verify this equation, finite element analysis is carried out on the model with vertical degree $a = 8°$, as shown in Fig. 3.19. According to Fig. 3.19, we can see that the hole stress concentration coefficient K_T is 2.69. Combined with the nominal stress method, we can see that the fatigue life prediction value of the model is 200,085 cycles. According to Eq. (3.3), we can see that the fatigue life prediction value of the model is 203,829 cycles. The relative error is about 2%. Therefore, it can be considered that it is proper to describe the effect of verticality on the fatigue life of the openings by the empirical Eq. (3.3).

3.2.3 Cylindricity

In the production practice, it has also been found that the diameter of the hole circle on drilling-in surface is slightly larger than the one on the drilling-out surface due to the swing of the arbor or other reasons, The actual hole looks more like a conical section. This tapered hole will not make full contact between the hole and the

Table 3.10 Tolerance of cylinder for each model

Model code	Cylindrical tolerance *c*/mm
C1	0.05
C2	0.1
C3	0.2
C4	0.4
C5	0.6
C6	0.8
C7	1.0
C8	1.2
C9	1.4

connecting bolt, but a certain gap. So stress concentration will occur on local point of contact, resulting in fatigue life reduction of the hole [3].

In this section, the cylindrical tolerance c is defined as the difference between the radius $R1$ of the orifice on the drilling surface and the radius $R2$ of the orifice on the other drilling surface which is,

$$c = R1 - R2 \tag{3.4}$$

In order to study the effect of cylindrical degree on the stress distribution of the hole, nine models are analyzed here under the condition that the diameter of the hole was kept constant on the drilling-in surface of all the holes. The cylindrical tolerances for each model are shown in Table 3.10.

The symmetrical constraint U1 = UR2 = 0 in direction 1 is applied to the left end of the model. The symmetrical constraint U2 = UR1 = 0 in direction 1 is applied to the bottom of the model. The symmetrical tension of right end is 100 N/mm.

After finite element analysis, the stress distribution and the corresponding stress concentration coefficient of opening with different cylindrical degrees can be obtained. Combined with the nominal stress method, the fatigue life of each model is shown in Table 3.11.

According to Table 3.11, it can be seen that the smaller the cylindrical tolerance is, the greater the fatigue life of the opening model is. The linear regression method was used to establish the empirical equation between cylindrical degree and hole fatigue life:

$$N = N_0 * e^{-0.21c} \tag{3.5}$$

In Eq. (3.5), c is the cylindrical tolerance of the opening, N is the fatigue life of the hole, and N_0 is the fatigue life of the ideal smooth opening.

In order to verify this equation, finite element analysis is carried out on the model with cylindrical degree $c = 0.3$ mm, as shown in Fig. 3.20. According to Fig. 3.20, we can see that the hole stress concentration coefficient K_T is 2.65. Combined with the nominal stress method, the fatigue life prediction value of the

Table 3.11 Fatigue life of different cylindrical models

Model code	Cylindrical tolerance *c*/mm	Stress concentration factor K_T	Fatigue life *N*/cycle
C1	0.05	2.58	225,111
C2	0.1	2.61	218,678
C3	0.2	2.64	212,362
C4	0.4	2.67	206,164
C5	0.6	2.71	198,016
C6	0.8	2.75	190,219
C7	1.0	2.79	182,564
C8	1.2	2.83	175,120
C9	1.4	2.86	169,675

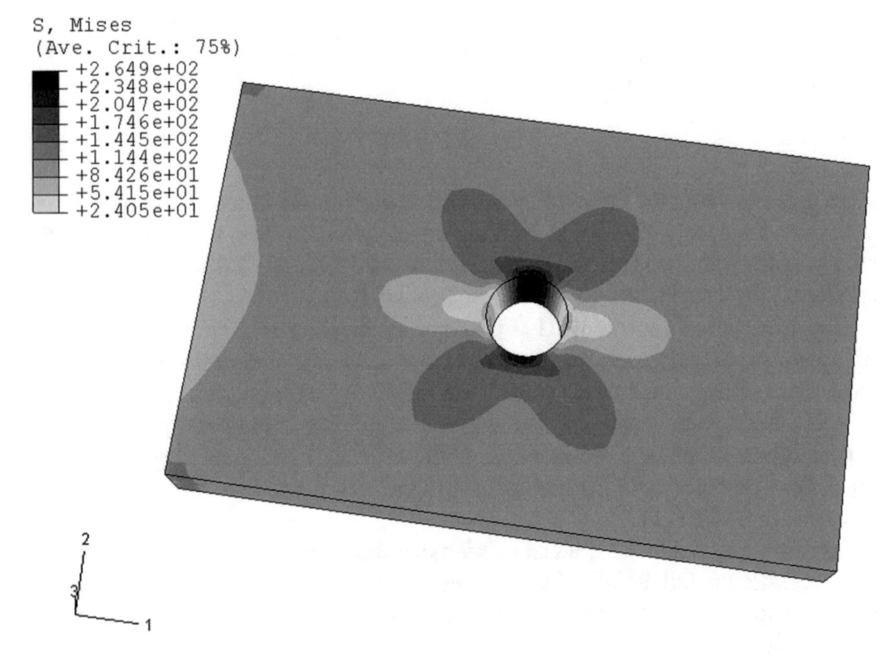

Fig. 3.20 Stress cloud of model with the cylindrical $c = 0.3$ mm

model is 210,283. According to Eq. (3.5), we can see that the fatigue life prediction value of the model is 211,089 cycles. The relative error is about 1%. Therefore, it can be considered that it is proper to describe the effect of cylindricity on the fatigue life of the openings by the empirical Eq. (3.5).

3.2.4 Roundness

In practice, due to the hole-making process, tool wear or other reasons, the true contour of the hole section is not an ideal circle, but with a certain shape changes, for example, a slightly triangular hole in reference [3]. The roundness of the hole is generally used to define this characteristic of the hole.

The roundness in the engineering survey is defined as an indicator used to limit the actual circle to its ideal circle variation. It is a requirement for the shape accuracy of the cross section of the cylindrical surface (conical surface) and any cross section of the sphere passing through the sphere center. The roundness tolerance zone refers to the area between the two concentric circles with the radius difference of the tolerance value t in the same normal section. Circle and tolerance are shown in Fig. 3.21 [15].

Due to the irregular shape of the circle, the stress concentration is more likely to occur at edge of the hole, which increases the fatigue crack source of the component and reduces the fatigue life of the hole.

Even in the same roundness tolerance, the contours of the real holes will vary widely and it is difficult to fully describe them with a definite mathematical modeling method. Some assumptions must be made to quantitatively analyze the effect of roundness on the fatigue life of the hole. In practice, the contours of some real holes can be approximated to polygons. And if the machining quality is poor, the contours are approximately triangular, as shown in Fig. 3.22.

Thus, the contour of the real hole is assumed to be a regular polygon in this chapter, and the difference in radius between the inscribed circle and the circumcircle of the polygon represents the roundness tolerance of the hole, as shown in Fig. 3.23. When the radius of the inscribed circle remains constant, the more polygonal dimension is, the smaller the roundness tolerance is and the closer to the inscribed circle the polygon is. Based on this assumption, the finite element method can be used to analyze the stress field of openings with different roundness tolerances.

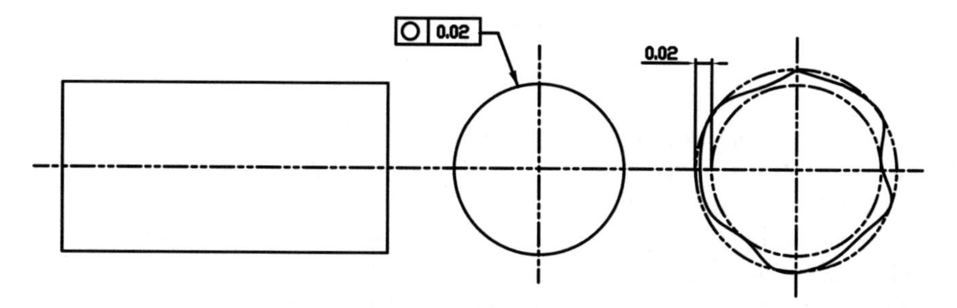

Fig. 3.21 Schematic diagram of roundness and tolerance

Fig. 3.22 Holes with the contour of approximate triangle

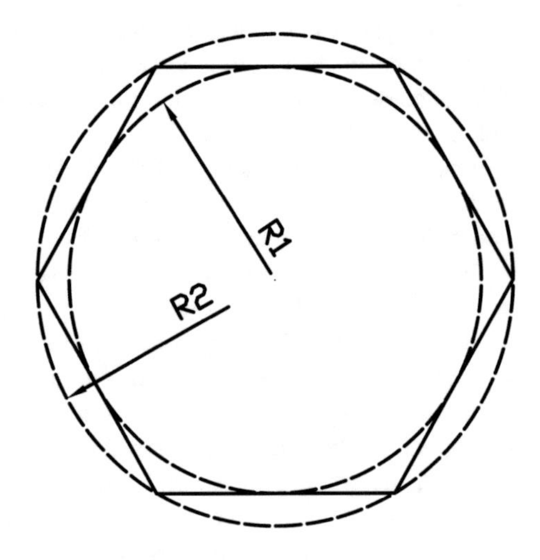

Fig. 3.23 Schematic diagram of the roundness of polygons

Table 3.12 Roundness tolerances for each model

Model code	Polygon dimension	Roundness tolerance *t*/mm
T1	16	0.05
T2	12	0.09
T3	10	0.13
T4	8	0.21
T5	7	0.27
T6	6	0.39

In order to study the effect of roundness tolerance on the hole stress field, six cases are analyzed in the case with the constant radius of the inscribed circle of the polygon. The roundness tolerance of each model is shown in Table 3.12.

The symmetrical constraint U1 = UR2 = 0 in direction 1 is applied to the left end of the model. The symmetrical constraint U2 = UR1 = 0 in direction 2 is applied to the bottom of the model. The symmetrical tension of right end of the model is 100 N/mm.

Table 3.13 Fatigue life of different roundness openings

Model code	Roundness tolerance t/ mm	Stress concentration factor K_T	Fatigue life N/ cycle
T1	0.05	2.67	214,454
T2	0.09	2.77	186,365
T3	0.13	2.84	171,477
T4	0.21	3.02	141,042
T5	0.27	3.15	123,050
T6	0.39	3.3	103,030

After finite element analysis, the stress distribution of the openings with different roundness and the corresponding stress concentration factor can be obtained. Combined with the nominal stress method, the fatigue life of each model is shown in Table 3.13.

According to Table 3.13, it can be seen that the smaller the roundness tolerance is, the greater the fatigue life of the open-cell model is. The linear regression method is used to establish the empirical equation between roundness and hole fatigue life:

$$N = N_0 * e^{-2.27t} \tag{3.6}$$

In Eq. (3.6), t is the roundness tolerance of the opening and N is the fatigue life of the hole; N_0 is the fatigue life of the ideal smooth opening.

The finite element analysis of the model with roundness t = 0.16 mm (positive 9 shape) is shown in Fig. 3.24 to verify this equation. According to Fig. 3.24, we can see that the hole stress concentration coefficient K_T is 2.96. Combined with the nominal stress method, the fatigue life prediction of the model is 149,075 cycles. According to Eq. (3.6), we can see that the fatigue life prediction value of the model is 161,110 cycles. The relative error is about 8%. Therefore, it can be considered that it is proper to describe the effect of roundness on the fatigue life of the openings by the empirical Eq. (3.6).

3.2.5 *Manufacturing Quality of Fasten Holes and Empirical Equation*

In practice, roughness, verticality, cylindricality, roundness of holes are not independent. Equations (3.2)–(3.6) are not independent either. Therefore, it is possible to approximate the relationship between these parameters of the holes and the fatigue life of the holes.

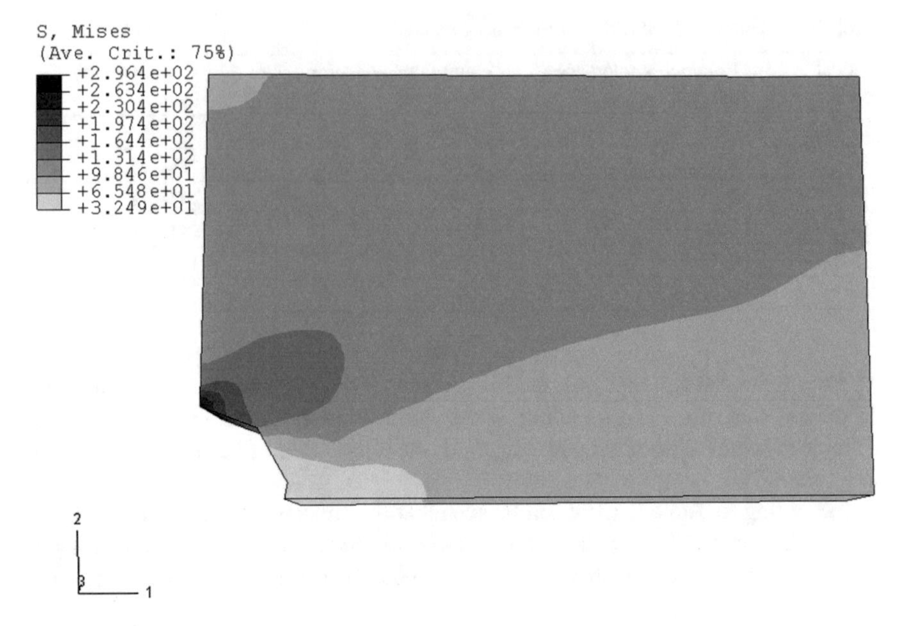

Fig. 3.24 Stress cloud of model with circularity t = 0.16 mm

$$N = N_0 * e^{(-0.136R - 0.016a - 0.21c - 2.27t)} \quad (3.7)$$

In Eq. (3.7), N is the fatigue life of the hole, N_0 is the fatigue life of the ideal smooth opening, R is the roughness of the opening, a is the angle between the hole axis and the surface normal, c is the cylindrical tolerance of the hole and t is the roundness tolerance of the opening.

3.3 Effect of Drilling on Fatigue Performance of Open Holes

The traditional hole-making is currently widely used in aircraft structure anti-fatigue fracture design. However, with the higher requirements of aircraft assembly line to technical of hole-making, the traditional multistep hole technology is difficult to ensure that the fatigue performance of the fastening hole meets the actual flight needs of the aircraft and thus advanced multi-step hole technology is brought in from abroad. Traditional hole-making process and the multi-step hole-making process is compared in Table 3.14. Traditional system of hole-making is characterized by multi-step processes, generally drilling, chambering, the first reaming, the second reaming and the third reaming or fourth reaming on important parts. The main process features of the multi-step hole-making technology, which is

Table 3.14 Two different pore-making processes

Name of process in hole-making		Traditional hole-making	Multi-step hole-making
Method code		DB	W
Main process		1 drill, 2 expand, 3 hinge	Compound of drilling, expansion, and hinge
Cutting amount	Rotating speed $r \cdot min^{-1}$	2300, 2150, 2150, 1400, 1400, 1400	3600
	Feed rate $mm \cdot r^{-1}$	0.04, 0.04, 0.04, 0.1, 0.1, 0.1	0.05
	Cutting speed $m \cdot min^{-1}$	34.3	88.2
Coolant		no	T80292B

commonly used in American aircraft factory assembly line, are drilling, expanding and hinging, and the drilling speed and rotate speed are guaranteed by tool.

In the traditional system of hole-making technology, the quality of the process is to a large extent constrained by the operator's technical level, as well as operating habits, emotions, and responsibility. If make holes in the assembly frame, there will be more prominent problems. In this section, the effects of the two kinds of pore-forming processes on the fatigue life are analyzed, based on samples with single dog bone holes.

3.3.1 Specimen Size and Loading

Material of single dog bone hole sample is 7050, with a total number of 42 samples, among which 21 is under the traditional system of hole-making and the other 21 is under multi-step process of the hole-making. The sample size is shown in Fig. 3.25.

The test is carried out on the test machine Instron 880, controlled by load. The fatigue load is sine wave cyclic load. The nominal stress levels are 80, 90, and 100 MPa, respectively. The stress ratio R is 0.06. According to the size of the load and the hole-making process, the samples are divided into six groups, and each contains seven.

3.3.2 Fatigue Test Results

Table 3.15 shows the fatigue test results of each group.

It can be seen from Table 3.15 that under the maximum load of 80 MPa, the life under the multi-step process is 2.1 times higher than that under the traditional hole-making process. Under the maximum load of 90 MPa, the life under

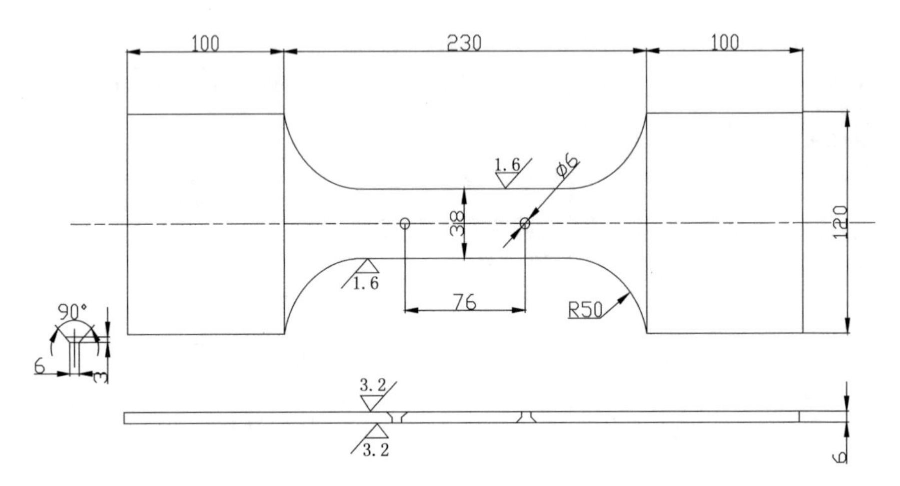

Fig. 3.25 Size of single bone sample

Table 3.15 Fatigue life of open hole in single canine under different pore-making and load

Processing technology	Maximum stress/ MPa	Mean life/ cycle	Coefficient of variation
Traditional hole-making DB	80	74,828	0.1193
Multi-step hole-making *W*	80	162,085	0.4111
Traditional hole-making DB	90	47,594	0.1051
Multi-step hole-making *W*	90	67,904	0.0991
Traditional hole-making DB	100	42,856	0.3191
Multi-step hole-making *W*	100	49,812	0.0837

multi-step increases by 1.4 times. Under the maximum load of 100 MPa, life under multi-step process increases by 1.2 times compared with the traditional hole-making process. It is shown that the multi-step hole-making process can significantly improve the fatigue life of the hole compared with the traditional hole-making process. The fatigue life is prolonged in different multiples under different stress levels. If the maximum load is smaller, the life expectancy is more obvious. In addition, in general, the life dispersibility under multi-step hole-making process is lower than that under the conventional pore-forming process.

References

1. Wang, Min. 1999. *Principle and Technology of Anti-fatigue Manufacture*. Jiangsu: Jiangsu Science and Technology Press.
2. Huang, Hongfa, Hai Yan, and Chunhu Tao. 1998. Analysis of Two Materials Defect Distribution. *Mechanic Strengh* 8: 237–239.
3. Carter Ralph, W., W. Steven Johnson, Paul Toivonen, Andrew Makeev, and J.C. Newman Jr. 2006. Effect of Various Aircraft Production Drilling Procedures on Hole Quality. *International Journal of Fatigue* 28: 943–950.
4. Pei, Xuming, Wuyi Chen, Bingyi Ren, and Yu. Han. 2001. Effects of Process on 7075 Aluminum Fastening Hole Surface and Structure. *China Nonferrous Metals Journal* 11 (4): 655–659.
5. ToParli, M., A. Ozel, and T. Aksoy. 1997. Effect of the Residual Stress on the Fatigue Crack Growth Behavior at Fastener Holes. *Materials Science and Engineering* A225: 196–203.
6. Yang, J.M., Y.C. Her, N.L. Han, et al. 2001. Laser Shock Peening on Fatigue Behavior of 2024-T3 Al Alloy with Fastener Holes and Stopholes. *Materials Science and Engineering* A298: 296–299.
7. Yao, Weixing. 2003. *Analysis of Structure Fatigue Life*. National Defense Industry Press.
8. Biswas, D.K., M. Venkatraman, C.S. Narendranath, and U.K. Chatterjee. 1992. Influence of Sulphide Inclusion on Ductility and Fracture Behaviour of Resulfurized HY-80 Steel. *Metallurgical Transactions A* 23: 1479–1492.
9. Ervasti, Esa, and Ulf Stahlberg. 2005. Void Initiation Close to a Macro-inclusion During Single Pass Reductions in the Hot Rolling of Steel Slabs: A Numerical Study. *Journal of Materials Processing Technology* 170: 142–150.
10. Wang, Wanpeng, Zhufeng Yue, and GuoZhi Yang. 2004. Damage Analysis of Tensile Sample Made of Metallic Material with Inclusions. *China Nonferrous Metals Journal* 14 (6): 949–955.
11. Dong, Dengke, Junyang Wang, and Fanjie Kong. 2000. Fatigue Quality Control of Fastening Hole and Study on Hole Making. *Mechanic Strengh* 22 (3): 214–216.
12. Cao, Changnian, Zhizhi Wang, and Xuanmin Zhao. 2000. Assessment and Conformance of Fastening Hole Fatigue Quality. *Northwestern Polytechnical University Journal* 18 (1): 15–18.
13. Songqing, Wu. 1990. *Application Guide of Surface Roughness*. Beijing: China Machine Press.
14. Zhang, Dongchu, and Xuming Pei. 2003. Effects of Process on Surface Roughness and Fatigue Life. *China Mechanical Engneering* 16: 1374–1377.
15. Kong, Qinghua, and Chuanshao Liu. 2002. *Basis of Limit Fits and Measuring Technique*. Shanghai: Tongji University Press.

Chapter 4
Anti-fatigue Strengthening Technology of Holes

In order to improve the resistance of the hole in the body structure, to make up for the defects caused by the drilling process, and to ensure the safety of the service life of the aircraft, people often use cold expansion, impression and hammering, and other anti-fatigue technology to further improve the fatigue resistance of the hole. The strengthening mechanism is mainly to introduce the residual compressive stress, reduce the stress peak, or average stress of the hole edge to achieve the purpose of improving its fatigue life [1]. In this chapter, the effects of cold expansion, impression and hammer strengthening techniques on the residual stress distribution, and the fatigue life of the hole are studied by means of finite element analysis and related experimental techniques.

4.1 Effect of Cold Expansion on Fatigue Performance of Open Holes

Cold expansion means that at room temperature, using the expansion tool that is more than the extruded material hardness, pressure is applied to the surface of the hole wall to deform the surface layer metal of the extruded part to form a residual compressive stress layer, to change the structure of the reinforcement layer, to increase the dislocation density, and to improve the surface quality of the hole wall. A variety of strengthening mechanisms work together to extend the crack initiation life and crack life, and reduce the crack speed to achieve the purpose of improving fatigue life, and cold expansion technology to increase the fatigue life at the same time, but also improve the structural hole resistance to stress corrosion and corrosion fatigue [2].

The cold expansion process has become one of the most popular anti-fatigue manufacturing technologies used in the aerospace industry. For example, Boeing has adopted the expansion strengthening method to increase the fatigue life in a

J. Liu et al., *Long-Life Design and Test Technology of Typical Aircraft Structures*,
https://doi.org/10.1007/978-981-10-8399-0_4

large number of dangerous nail holes in its aircraft structure. The same method to improve nail hole fatigue life has been used in China. Chengdu Aircraft Industry Company on a plane with nearly 4,000 holes uses a cold expansion process; the material has high-strength steel (such as 30CrMnSiNi2A), titanium and aluminum alloy. The diameter of the expanded hole is in the range of 4–70 mm.

Fastening of the cold expansion technology in the aviation industry has been applied for more than 60 years; this technology is to improve fatigue life without increasing the weight of the effective way. To sum up, the uses of local surface plastic deformation to improving fatigue life by the main methods are rolling polishing, ball pressure treatment, direct mandrel expansion, sleeve expansion strengthening process, casing expansion, and so on. The main difference between them is the shape of the extruded object and the use of sleeves in cold expansion [3]. The use of sleeves is mainly to avoid damage and protrusion of hole caused by friction and interference during expansion process, and can make the hole residual stress distribution more uniform. However, using cold expansion with sleeves, convex is formed on hold wall. Convex must be removed by the subsequent reaming. The whole working process is very complex. In addition, the cost is high for reason that the expanded sleeve could not be used in next expansion. The direct squeezing of the mandrel causes the hole wall to have slight axial abrasion and the small angle of the hole angle, but the process cost is simple and low cost. At present, both at inland and abroad, two cold expansion processes are used [4].

4.1.1 Direct Mandrel Expansion Process and Its Parameters

Direct mandrel expansion is developed in the 1950s in the hole cold expansion process in the most widely used, the most mature of a technology. Douglas is the first to use this technology to improve the fatigue life of aircraft aluminum holes. This method is to use milling or broaching to prefabricate an initial hole, with a fully lubricated mandrel in the form of interference forced through the initial hole. In order to eliminate the effect of roughness on fatigue life, after cold expansion, the hole is machined to the final size with a reamer. This method is simple to use tools and processes, suitable for low-interference cold expansion (2–6%) [5, 6]. The relative expansion is defined as the ratio of the absolute expansion to the diameter of the primary hole:

$$\mathrm{Er} = \frac{d_0 - D_0}{D_0} \times 100\% \tag{4.1}$$

In Eq. (4.1), Er is the relative expansion amount; d_0 is the diameter of the mandrel; and D_0 is the initial hole diameter.

In general, the surface of the specimen where the mandrel enters the hole is called the entrance face. The surface of specimen where the mandrel leaves the hole is called the exit face. Figure 4.1 shows the direct mandrel expansion.

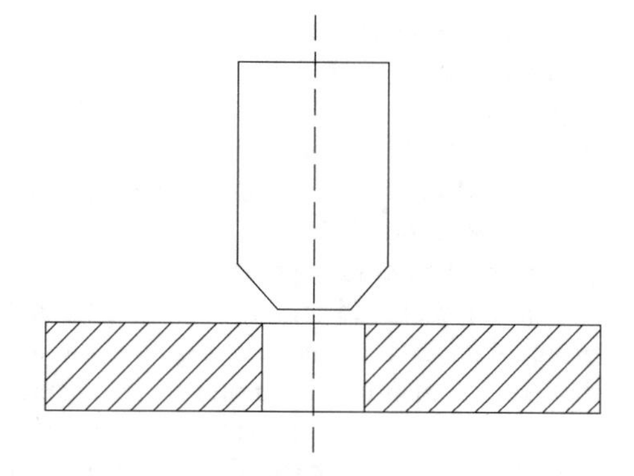

Fig. 4.1 Schematic diagram of direct mandrel expansion

This section analyzes the effects of four different cold expansion capacities: 0 (non-cold expansion), 2, 4, and 6% on the fatigue life of the open holes fatigue test specimens are cut from 4mm thickness 2A12 aluminum plate in rolling direction. The final size data is shown in Fig. 4.2. The stress–strain relationship of the 2A12-T4 aluminum plate is shown in Fig. 4.3. The elastic modulus E = 68 GPa, the yield stress $\sigma_b = 320$ MPa, $\sigma_s = 430$ MPa, and Poisson's ratio μ = 0.33.

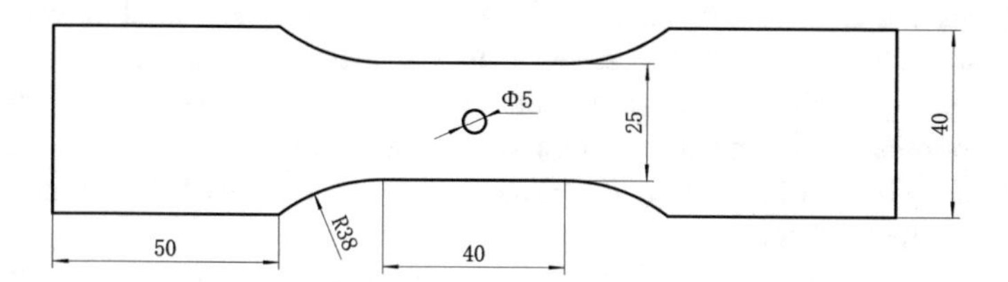

Fig. 4.2 Specimen size

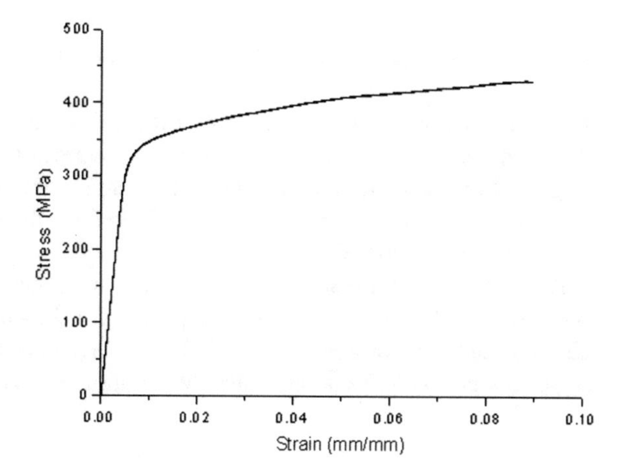

Fig. 4.3 Stress–strain curves of 2A12-T4

Four groups of specimens containing holes were prepared and each group has seven specimens. The first group is the holes left with no cold expansion. The cold expansion was applied to the other three groups of specimens by pushing a pre-lubricated hardened steel mandrel through the pre-drilled holes without sleeves. To support the fatigue specimen during cold expansion, a 20-mm-thick steel plate with center hole of 20 mm was placed on reverse side of it. The second group of specimens had cold expanded with pre-drilled hole of 4.95 mm, the largest diameter of mandrel 5.05 mm, and the interference on holes diameter was 2%. The third group of specimens had cold expanded with pre-drilled hole of 4.9 mm, the largest diameter of mandrel 5.1 mm, and the interference on holes diameter was 4%. The fourth group of specimens had cold expanded with pre-drilled hole of 4.8 mm, the largest diameter of mandrel 5.1 mm, and the interference on holes diameter was 6%. After cold expansion, the three sets of cold expansion holes are machined to a final size with a Φ5 mm reamer.

4.1.2 Residual Stress Measurements

It is well known that residual stresses around cold expanded holes are dependent on an angular, radial, and the thickness position. There is a significant difference in the tangential compressive residual stresses on the mandrel entrance and exit face. Crack initiating at the corner of the hole and entrance face grows the fastest and become dominant. So the distribution of residual stress has the main effect on fatigue life. Thus, in this paper, X-ray measurements were used to obtain residual stresses on the entrance face.

The X-ray machine (MSF 2905) was used to measure surface residual stress of the cold expanded holes. The beam size of X-ray is 1×1 mm and the error of this X-ray machine is ± 20 MPa (Fig. 4.4).

4.1.2.1 Tangential Residual Stress Distribution

The residual stress of the fatigue life of the specimen is mainly tangential residual stress, and after the cold expansion, the test specimen is subjected to the test of the tangential residual stress σ_θ. For the cold reaming, fatigue cracking usually occurs at the extruded end of the hole [7, 8]. Therefore, the surface tangential residual stress σ_θ is measured on the extruded surface of the opening. The cold expansion volume of 2, 4, and 6% of the test groups were randomly taken out of a hole tangential residual stress test. The test position is shown in points 1–4 as shown in Fig. 4.5, and the test area is 1 mm^2. The test results are shown in Table 4.1 (the negative sign in the Table indicates the compressive stress).

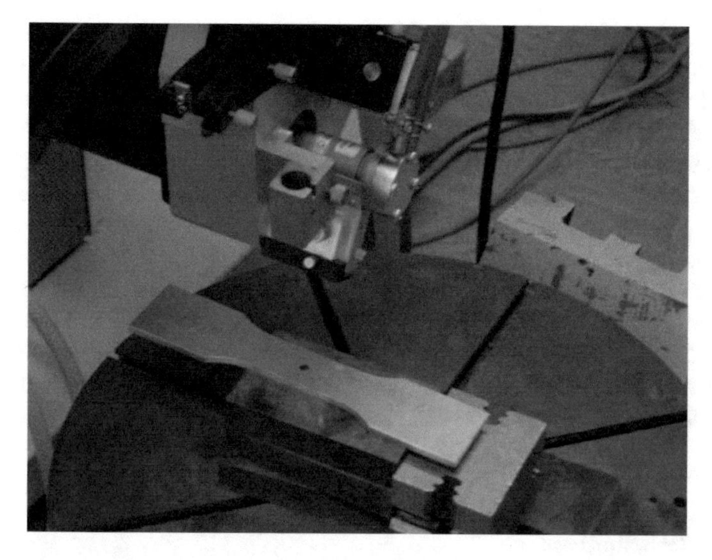

Fig. 4.4 MSF-2905 X-ray residual stress testing apparatus

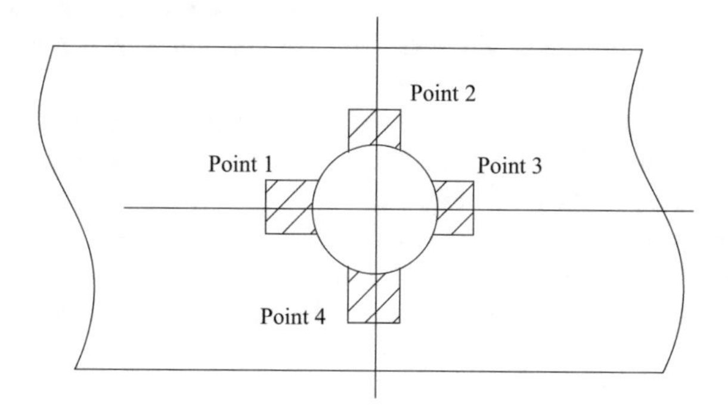

Fig. 4.5 Schematic diagram of the residual stress measurement of the orifice

Table 4.1 Surface residual stress test results (MPa)

Cold expansion (%)	point 1	point 2	point 3	point 4
2	−98	−32	−56	−37
4	−142	−50	−120	−26
6	−132	−99	−168	−39

As can be seen from Table 4.1, the tangential compressive stress of the specimen with 6% cold expansion is the largest. In addition, the residual stress values of point 1 and point 3, point 2, and point 4 of each specimen are different, which indicates that the distribution of tangential residual stress on the orifice expansion surface is asymmetric.

The asymmetric phenomenon of the residual stress in the orifice can be explained by the process of the direct cold expansion process. First, when the mandrel is extruded, the core axis is aligned with the hole. The central axis is not completely coincident, but there is a certain angle; this error will lead to the hole and the mandrel in the hole at different locations of the interference is not exactly the same, so the hole plastic area is inconsistent. The distribution of the residual stress in the orifice is naturally asymmetrical. Second, after the mandrel is extruded, it is necessary to use the reamer to hinge the hole to the final size. Due to the vibration of the tool or other reasons, the different positions of the hole of the cutting amount are different; therefore, due to the role of the hole caused by the residual stress, release is also different; after cutting, the residual stress distribution of the orifice will be asymmetric. Due to these two main reasons, the residual stress distribution of the orifice of the cold expansion hole is asymmetric.

4.1.2.2 Tangential Residual Stress in Longitudinal and Transverse Distributions

In order to study the change of the tangential residual stress at different positions of the orifice, the tangential residual stress is tested with 6% cold expansion. The test direction is longitudinal and transverse. The test position is shown in Fig. 4.6, and the pitch of each measurement point is 1 mm.

After the test, the tangential residual stress distribution in the longitudinal direction is shown in Fig. 4.7. The tangential residual stress distribution in the transverse direction is shown in Fig. 4.8.

As can be seen from the comparison of Figs. 4.7 and 4.8, the tangential compressive residual stress region is about 3 mm in the longitudinal direction and about 3.5 mm in the transverse direction. Due to the influence of the aspect ratio of the

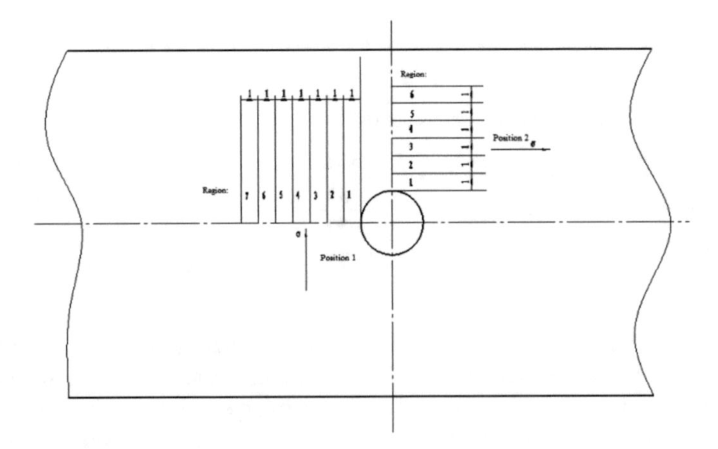

Fig. 4.6 Specimen test position diagram

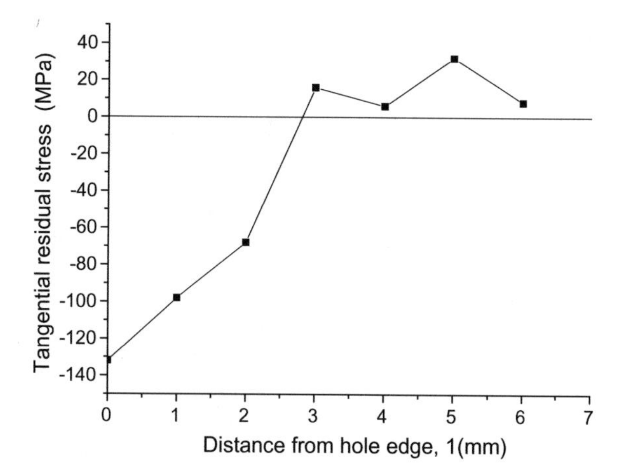

Fig. 4.7 Hole tangential residual stress distribution in the longitudinal direction

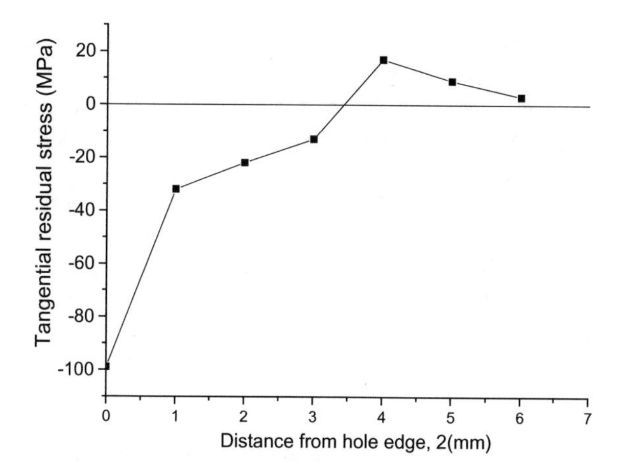

Fig. 4.8 Distribution of residual stress in the transverse direction of the hole tangential

specimen, the tangential compressive residual stress region in the transverse direction is larger than the tangential compressive residual stress region in the longitudinal direction. In addition, the tangential compressive residual stress value at the orifice is the largest, and the tangential residual compressive stress decays in both directions.

4.1.2.3 Layered Depth Distribution of Tangential Residual Stress

In order to test the distribution of the tangential residual stress in the depth direction, the specimen is subjected to chemical corrosion and delamination. The composition of the chemical etchant is HF 10 ml, HCL 5 ml, HNO_3 5 ml, and H_2O 100 ml. The etch time of each layer is 15 min. The test position is point 3 as shown in Fig. 4.5; the corrosion area is 5 × 5 mm^2; the test area is 1 × 1 mm^2; and the

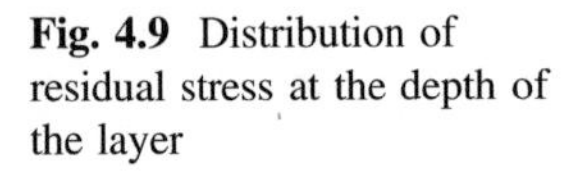

Fig. 4.9 Distribution of residual stress at the depth of the layer

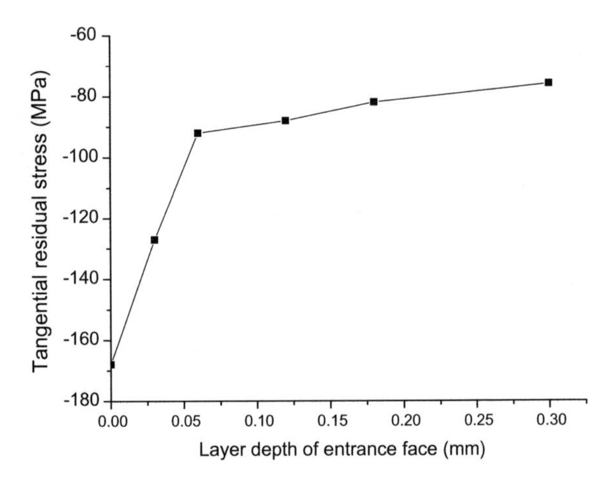

specimen with 6% cold expansion is selected for layer corrosion. The total depth of corrosion is 0.3 mm. Because the corrosion time is short, corrosion caused by the release of residual stress is small, which can be ignored. The test results of the tangential residual stress in the deep direction of the squeeze surface are shown in Fig. 4.9.

It can be seen from Fig. 4.9 that the tangential compressive residual stress distribution is not uniform along layer depth of entrance face and the max tangential compressive residual stress occurred on top layer. The tangential residual stress decrease quickly along layer thickness direction. For instance, the value of residual stress at layer of 0.3 mm depth, is only 50% of top layer.

4.1.3 Fatigue Test Results and Fracture Analysis

All the specimens were tested on a servo-hydraulic fatigue testing machine (Instron 8802). The uniaxial fatigue tests were carried out using constant load amplitude and sine wave loads with $R = 0.1$, where R is the ratio of the minimum load to the maximum load. Frequency of cycling in all fatigue tests was 16 Hz. The maximum stress used in this study was 120 MPa.

After the experiment, the median fatigue life, standard deviation, and coefficient of variation of each group are shown in Table 4.2. The coefficient of variation reflects the dispersibility of the experimental data.

As can be seen from Table 4.2, the fatigue test yielded the result that with these dimensions; 6% diameter interference is best for cold expansion. The fatigue life of cold expansion hole is about six times that of non-cold expansion hole. Therefore, the 6% cold expansion volume is the optimum amount of expansion of the specimen. However, the cold expansion process to improve the life of the same time also led to increased dispersion of life.

Table 4.2 Mean fatigue life, standard deviation, and variation coefficient of specimens with different cold expansions

Cold expansion (%)	Mean fatigue life/cycle	Standard deviation	Variation coefficient
0	115,080	7,846.9038	0.068
2	186,446	45,005.9	0.24
4	566,356	54,022	0.09
6	645,645	283,482	0.44

After fatigue testing was completed, the fracture surfaces of specimens were observed by scanning electron microscope. As expected, since the cold expanded hole has significant compressive stressed down the bore of the hole, crack always initiates at entrance face. The spreading speed of the crack in the direction of the crowded surface is significantly larger than its spreading speed in the thickness direction. The typical crack pattern is shown in Fig. 4.10.

From the above finite element analysis, we can see that the tangential residual compressive stress on the squeeze surface is the smallest, which is the reason why the fatigue crack source mainly occurs in the squeeze surface. As the residual compressive stress decreases gradually along the direction of the tangential direction, the residual compressive stress decreases gradually along the thickness direction. Therefore, the resistance of the crack in the direction of the surface decreases gradually, and the resistance decreases gradually as the thickness direction increases. This is the main reason why the expansion speed in the crack surface direction is significantly larger than its expansion speed in the thickness direction.

Fig. 4.10 Typical crack of cold expansion holes

4.1.4 FEM Analysis

In order to compare with the test results of the residual stress test on the surface of the cold expansion hole, a finite element analysis model of cold expansion was established based on 6% of the specimen size data. Since the structure and load are symmetrical about 1-direction and 2-directions, only the 1/4 of the original model is analyzed. The material parameters of the mandrel are $E = 210$ GPa and $\mu = 0.3$ (Fig. 4.11).

The symmetry constraint condition is applied to the symmetry plane of the mandrel and the perforated aluminum plate. Since the exit face of the aluminum plate is pressed on the steel plate during cold expansion, the 3-direction restraining condition is applied to the bottom surface of the plate. The friction between the mandrel and the hole on the residual strain is very small, so the friction here is ignored [9]. The contact surface on the mandrel is defined as the contact master face, and the corresponding face of the hole is defined as the contact salve surface. The stress–strain relationship of the obtained aluminum plate is introduced in the calculated elastoplastic model.

Numerical simulation of tangential residual stress after cold expansion is shown in Fig. 4.12.

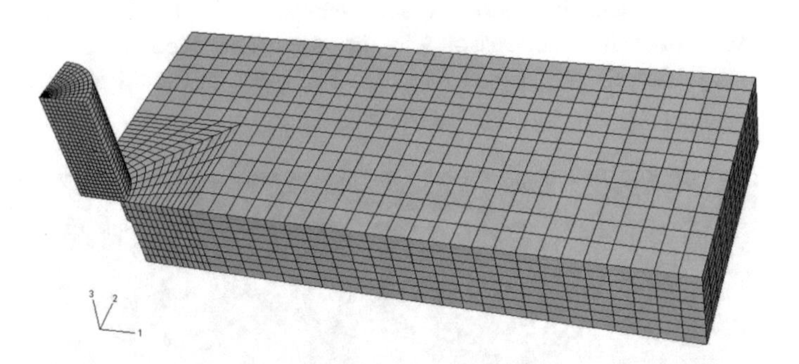

Fig. 4.11 Finite element model of cold expansion

Fig. 4.12 Tangent residual stress distribution of cold expansion holes

It can be seen from Fig. 4.12 that the tangential residual stress distribution of the hole wall is not uniform in the thickness direction. The maximum compressive residual stress occurs on the exit face with a value of about −480 MPa and the minimum compressive residual stress occurs in the entrance face. The value is about −300 MPa.

The tangential residual stresses are shown in the longitudinal and transverse directions on the entrance face, mid-plane, and exit face, respectively, as shown in Figs. 4.13 and 4.14.

As can be seen from Figs. 4.13 and 4.14, the length of the compressive residual stress zone in the longitudinal direction is about 3.3 mm and the length of the compressive residual stress zone in the transverse direction is about 3.5 mm. The compressive residual stress region in the transverse direction is larger than that in

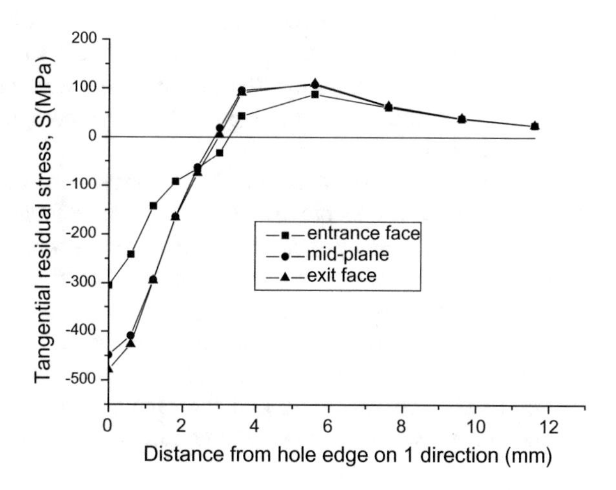

Fig. 4.13 Tangential residual stresses are distributed longitudinally

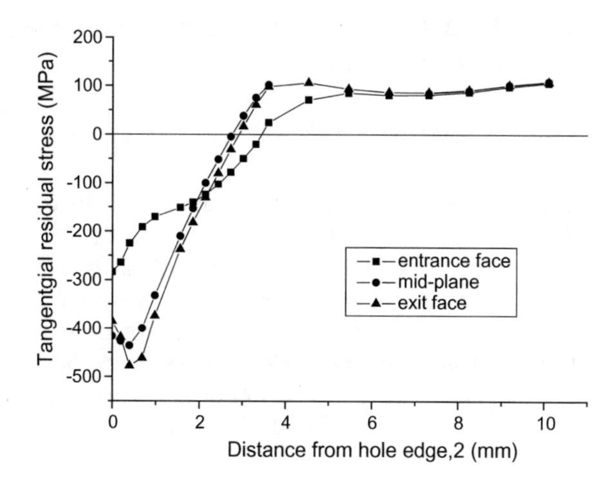

Fig. 4.14 Tangential residual stresses are distributed laterally along the hole

the longitudinal region. This conclusion agrees well with the experimental results, indicating that the finite element model is reasonable.

As can be seen from Fig. 4.13, in the longitudinal direction, the maximum tangential residual compressive stress at the three levels occurs at the orifice. In terms of numerical value, the exit face is the largest, the middle face is the second, and the entrance face is the smallest. With the distance from the orifice, the tangential residual compressive stress decays rapidly, the tangential residual compressive stress on the exit face is the fastest, and the tangential residual compressive stress on the entrance face is the slowest.

As can be seen from Fig. 4.14, in the horizontal, the three sides of the maximum tangential residual compressive stress generation site are inconsistent. On the entrance face, the maximum tangential residual compressive stress occurs at the orifice, while the maximum tangential residual compressive stress on the intermediate and exit face occurs at about 0.5 mm from the orifice. In terms of numerical values, the middle face is the largest, the exit face is the second, and the entrance face is the smallest. With the distance from the orifice, the tangential residual compressive stress decays rapidly, the tangential residual compressive stress on the exit face is the fastest, and the tangential residual compressive stress on the entrance face is the slowest.

In addition, the results obtained by the finite element analysis are compared with the experimental results obtained before. It can be found that the results obtained by the finite element analysis are larger than the experimental results. There are two main reasons for this difference. First, the actual orifice of the stress state and boundary conditions are more complex, which can be verified from the actual residual stress distribution. In the finite element analysis, the stress and boundary conditions of the orifice plate are idealized. Second, the hole is reamed after the mandrel has been squeezed, and the reaming process causes the release of the residual stress, while the finite element model ignores this effect when analyzed.

4.2 Effect of Impression on Fatigue Performance of Open Holes

4.2.1 Impression Process

Fillet impression strengthening is the use of certain rounded head with an arc groove at both ends of the fastener hole so that the hole circumference produces residual compressive stress, thereby increasing the fatigue life of the fastening holes. The advantage is that without the removal of structural materials in the case, the hole at both ends of the formation of a better situation of the arc-shaped chamfering improved the traditional use of mechanical methods to remove structural materials and the formation of stress more concentrated sharp angle down angle. The mechanism of residual compressive stress is mainly through the indenter

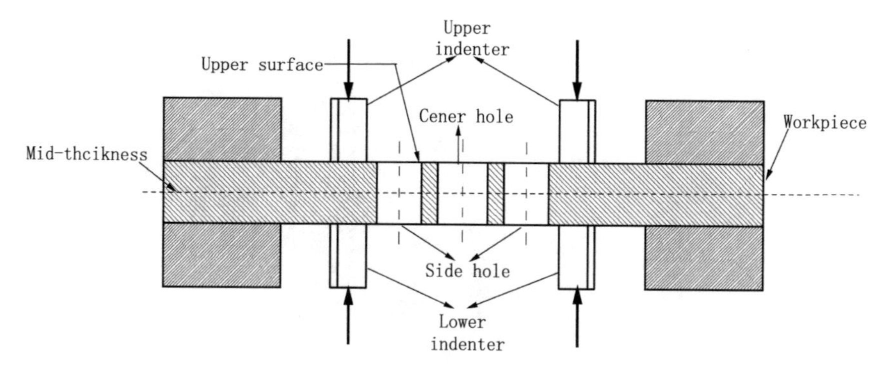

Fig. 4.15 Schematic diagram of rounded impression

on the thickness of the fasteners in a certain amount of expansion, so that the expansion contact surface and the hole were plastic deformation and elastic deformation. When the indenter is unloaded, the elastic deformation of the hole is to be restored, but the residual compressive stress is generated in the hole due to the failure of the plastic deformation of the contact surface and the recovery of the elastic deformation of the hole. The existing literature statistics show that the use of rounded embossed to strengthen the technology, and the aircraft structure of the junction of the fatigue life than the traditional method of chamfering increased by 2 times [10, 11]. The process of fillet impression is shown in Fig. 4.15.

4.2.2 Influence of Indenter Size on Residual Stress Distribution

The finite element model of the perforated specimen with thickness of 80 mm, width 40 mm, thickness of 5 mm, and aperture R of 5 mm was established, see Fig. 4.16. The material of the open-hole test plate is 2A12-T4, and its elastic-plasticity is shown in Fig. 4.3. The material of the indenter is 45 steel, which is regarded as rigid body for simplified calculation. A quarter of the model was analyzed, as a result of the geometric shape and the symmetry of the load.

In this section, the effects of different fillet radians (θ = 90°, 110°, 130°, 150°, 160°, and 170°) and relative radii (d = 1, 2, 3, 4, and 5 mm) on the residual stress distribution in the perforations are mainly considered, wherein the relative radius d of the indenter is defined as the radius of the indenter and the aperture radius difference. Indenter thickness of 1 mm and height of 5 mm remain unchanged, see Fig. 4.17.

The simulation process of surface impression is divided into two steps: pressing and unloading of the indenter.

Fixed and loading methods are as follows: (1) The indenter is only allowed to move in the Z direction, and the displacement load is applied to the indenter is

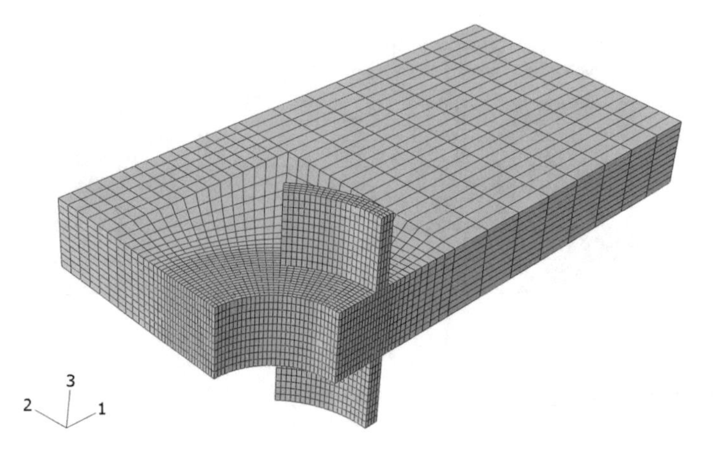

Fig. 4.16 Model grid diagram

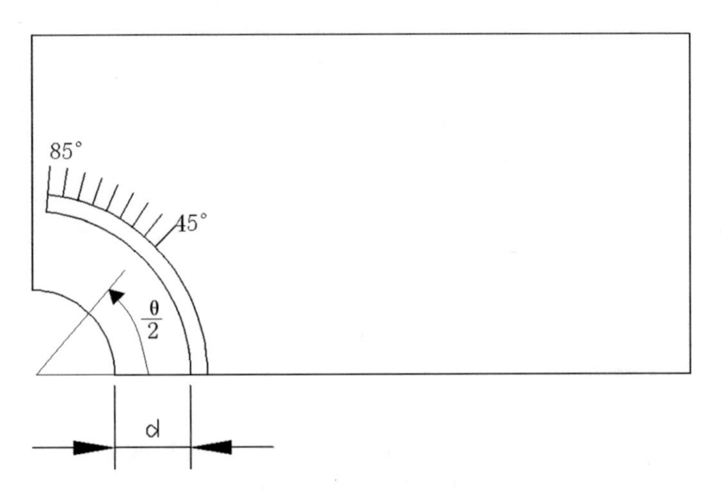

Fig. 4.17 Dimension optimization of 1/4 model indenter

0.1 mm, in which the upper and lower indenters are opposite; the surface impression process (loading and unloading at the same time) is carried out by displacement control; (2) As a result of the 1/4 model, the symmetrical boundary conditions are used on the two symmetry surfaces, respectively; and (3) apply the fixed confinement to the two clamping sections of the specimen.

Fillet impression can improve the fatigue life mainly by generating residual compressive stress at the hole, reducing the mean value of the fatigue load spectrum, while the residual compressive stress plays a major role in the circumferential (tangential) residual compressive stress. Figure 4.18 shows the circumferential residual compressive stress distribution of the orifice to the transverse edge under indentation enhancement at different relative radii and rounded radians.

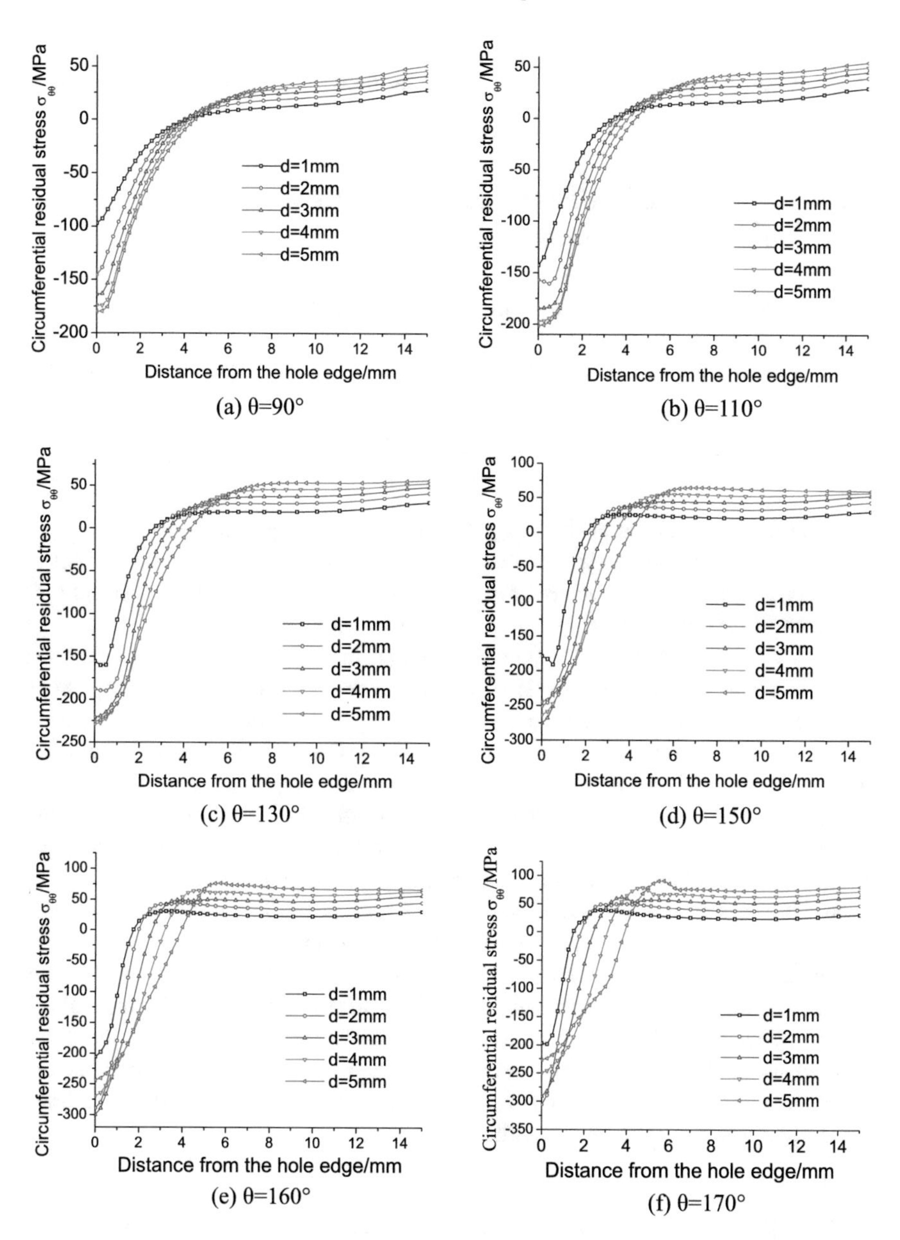

Fig. 4.18 Distribution of residual stress in the circumferential circumference of the indenter with different relative radii and radians

As can be seen from Fig. 4.18, in most cases, the maximum circumferential residual compressive stress appears in the hole wall. With the increase of curvature and relative radius, the maximum circumferential residual compressive stress and the maximum circumferential tensile stress increase. The size of the residual compressive stress is mainly determined by the relative radius, but when the relative radius is constant, the arc increases and the residual compressive stress increases.

4.2.3 Influence of Different Factors on Residual Stress of Impression-Reinforced Holes

4.2.3.1 Effect of Extrusion Amount on Residual Stress Distribution

The extrusion amount of the surface impression reinforcement i is determined by

$$i = \frac{h}{H/2} \times 100\% \tag{4.2}$$

where H is the thickness of the test piece and h is the expansion depth.

This section considers six cases of i = 1%, 2%, 3%, 4%, 5%, and 6%, respectively. Figure 4.19 shows the circumferential and radial residual stress distributions for different circumferential pressures. As can be seen from Fig. 4.19:

(a) For the circumferential residual stress $\sigma_{\theta\theta}$, in addition to 1% expansion, the circumferential residual compressive stress reaches a maximum at a very small distance from the edge of the hole. And after this point for the different expansions, volume has a tendency to decrease. With the increase of the expansion amount, the circumferential residual stress also increases, but the amplitude is getting smaller and smaller, and the residual stress in the circumferential wall does not increase until the expansion amount is 6%. The compressive stress zone of the residual stress in the circumferential direction is 12.5 mm from the hole edge.

(b) The radial residual stress σ_{rr} is the same as the circumferential residual pressure $\sigma_{\theta\theta}$, and after reaching a certain distance from the hole, it reaches its maximum and begins to decrease after this point. But the distance from the hole edge is greater than the circumferential residual pressure $\sigma_{\theta\theta}$ of the distance, about 2.5 mm. Unlike the residual pressure in the circumferential direction $\sigma_{\theta\theta}$, the radial residual stress of the wall is almost constant as the expansion amount increases, but the overall trend is increased. The compressive stress region of the radial residual stress is almost the same as the width of the specimen.

4.2.3.2 Effect of Friction Between Indenter and Specimen on Residual Stress Distribution

Figure 4.20 shows the distribution of circumferential and radial residual stresses on the circumferential and radial forces under different extrusion amount. It can be seen that the effect of friction on the residual stress distribution in the holes is subtle

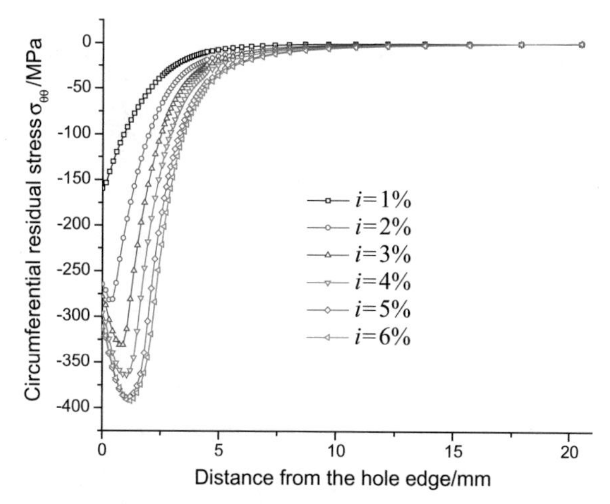

(a) Circumferential residual stress distribution

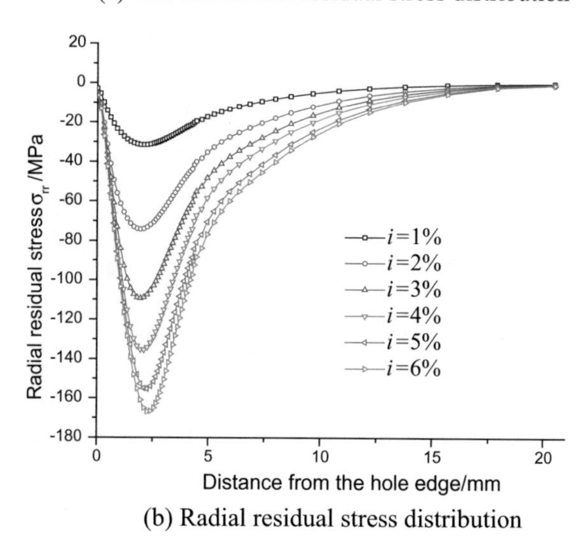

(b) Radial residual stress distribution

Fig. 4.19 Distribution of residual stresses in peripheral bars corresponding to different expansions

and local. When considering the frictional force between the indenter and the test piece, the maximum value of the circumferential residual compressive stress and the radial residual compressive stress increases. For the residual compressive stress in the circumferential direction, the friction force mainly affects the area from the hole edge to the circumferential residual stress to the maximum value, while the radial residual compressive stress mainly affects the region of the maximum residual compressive stress.

Fig. 4.20 Effect of friction on residual stress distribution in hole

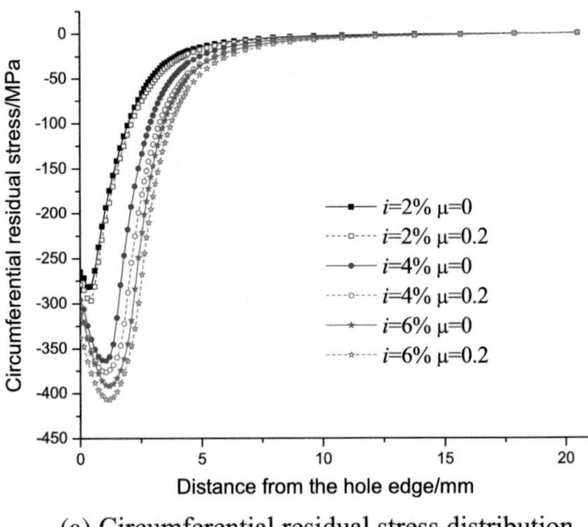

(a) Circumferential residual stress distribution

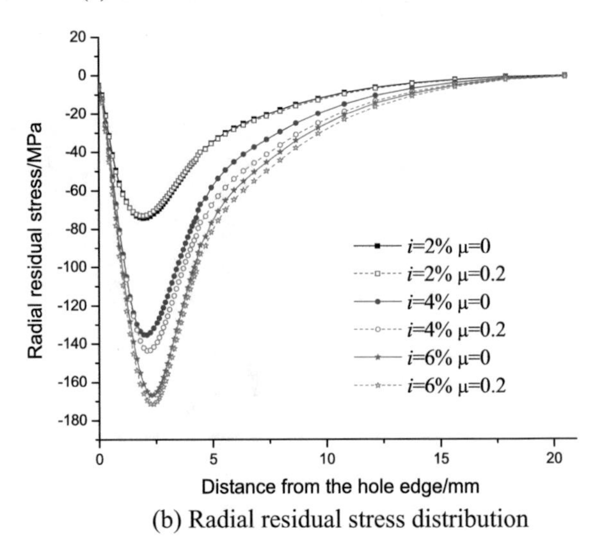

(b) Radial residual stress distribution

4.2.3.3 Influence of Loading Sequence on Residual Stress Distribution

In the actual project, the surface impression process may not be loading and unloading at the same time, but first load the specimen on the surface, and then unloading, and then load the specimen under the surface, and then unloading four stages. In view of this, in this section, using the model of extrusion amount $i = 4\%$ for example, it analyze the effect of two load sequence of indenters on residual stress distribution around hole. The first load sequence is that both the upper indenter and lower indenter load and unload at same time. The second load sequence is that upper indenter load and unload firstly, then the lower indenter load and unload.

Figure 4.21 shows the circumferential stress distribution in the thickness direction under different loading sequences. Compared with the numerical simulation results under the two loading sequences, the residual stress in the hole is symmetrical in the thickness direction when the upper and lower indenters are loaded at the same time. On the contrary, the upper and lower sequence loading, the hole residual stress is not symmetrical. Up and down at the same time, the maximum circumferential and radial residual stress occurs at 1/16 of the thickness and the upper and lower surfaces. While the sequential circumferential load and radial residual stress appear at the 1/4 and lower surfaces of the thickness. The minimum circumferential and radial residual stresses are present near the central plane.

Figure 4.22 shows the residual stress distribution from the orifice to the transverse side in different sections under different loading sequences. Comparing the results of the two models, the circumferential variation of the circumferential compressional compressive stress and the radial residual compressive stress in different sections is the same. Compared with the simultaneous loading and unloading of the upper and lower indenter, the circumferential residual compressive stress on the center plane increases from the hole edge to the maximum value at the time of sequential loading, but decreases on the upper and lower surfaces. And the circumferential residual compressive stress of the lower surface (second loading surface) is larger than the upper surface. And the radial residual stress is completely opposite to the circumferential residual stress. Contrast to simultaneous load, the radial residual compressive stress decreases from hole to the maximum stress point on central plane but increase on upper surface and lower surface under load sequence.

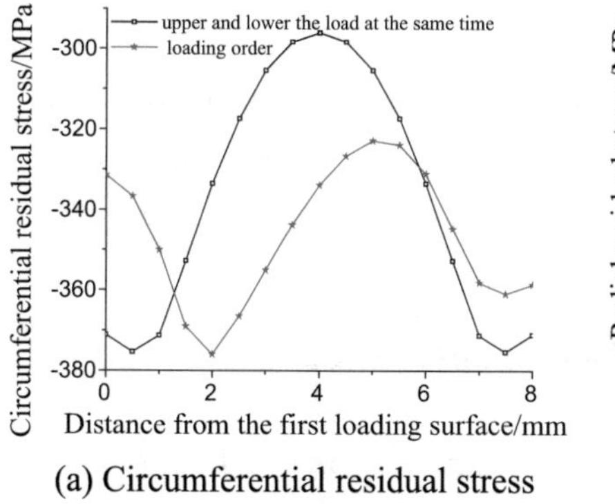

(a) Circumferential residual stress distribution

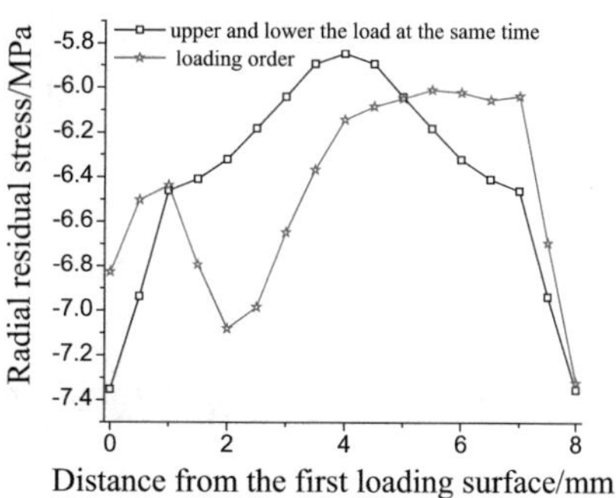

(b) Radial residual stress distribution

Fig. 4.21 Residual stress distribution of holes in different thicknesses in different loading orders ($i = 4\%$)

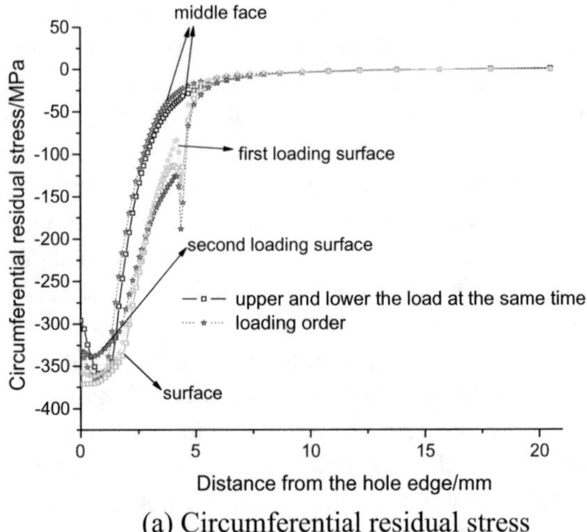

(a) Circumferential residual stress distribution

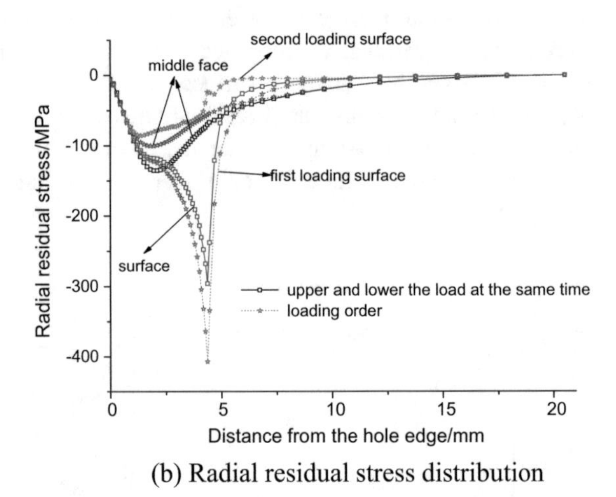

(b) Radial residual stress distribution

Fig. 4.22 Residual stress distribution in different cross sections of different loading sequences ($i = 4\%$)

4.2.4 Fatigue Test of Three-Hole Sample Impressed

The material of the three-hole impressed sample is 2A12-T4. The material parameters were measured by tensile test: $E = 68$ GPa, $\mu = 0.3$. The test piece is an 8-mm-thick plate with three holes on the loading axis. The geometry of the test specimen is shown in Fig. 4.23.

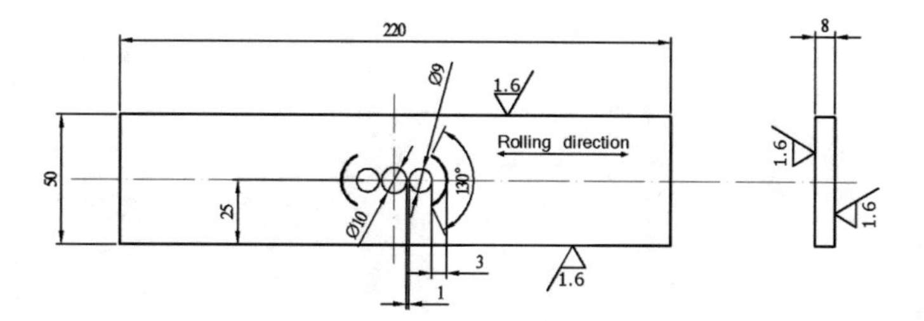

Fig. 4.23 Geometry of three-hole specimen

Table 4.3 Mean fatigue life, standard deviation, and coefficient of variation for different impressed depth specimens (σ_{max} = 200 MPa, R = 0.1)

Impressed depth/ mm	Mean fatigue life/ cycle	Standard deviation	Coefficient of variation
0	61,553	5,739.36	0.093
0.1	109,139	18,762.32	0.171
0.2	118,440	17,901.47	0.151
0.3	128,340	6,610.90	0.051

All specimens were tested on a servo-hydraulic fatigue tester (MTS810). Using the sine wave amplitude spectrum, the maximum nominal stress is σ_{max} = 200 MPa, the stress ratio is R = 0.1, the load frequency is 12 Hz. The specimens were divided into four groups according to the depth of impression, each group of eight pieces. The first group was unprinted original specimen, and the other three groups of impress depths of 0.1, 0.2, and 0.3 mm.

After the experiment, the median fatigue life, standard deviation, and coefficient of variation of each group are shown in Table 4.3. It can be seen that the greater the impress depth, the longer the fatigue life, and when the impress depth is 0.3 mm, the longest the life. At the same time as the depth of the impress increases, the dispersion of fatigue life is reduced.

Through the observation of the fracture of the specimen, it can be found that the fatigue crack of the embossed material initiated in the central plane of the middle hole, in the form of surface cracks, as shown in Fig. 4.24. The compressive stress caused by the indentation weakened the stress concentration of the hole, which hindered the occurrence of fatigue damage and delayed the crack propagation rate, thus improving the anti-fatigue performance of the fastening hole.

Fig. 4.24 Fatigue fracture of the impress enhancement hole

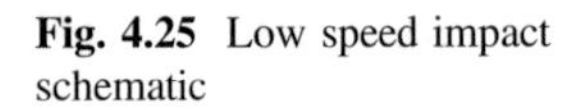
Fig. 4.25 Low speed impact schematic

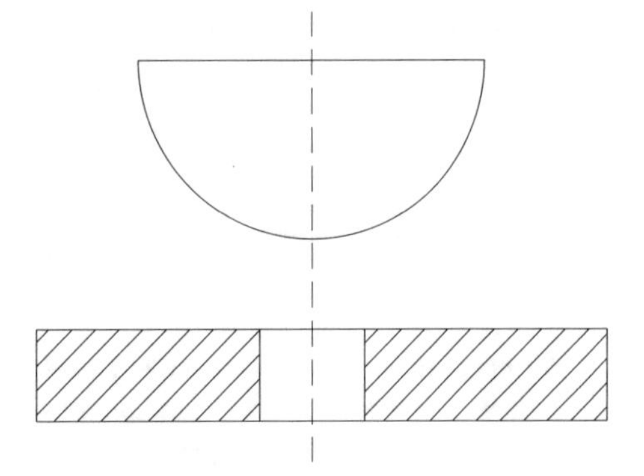

4.3 Effect of Hammering on Fatigue Performance of Open Hole

4.3.1 Hammering and Fatigue Testing

In this section, we study the effect of hammer energy on the fatigue properties of aluminum alloy under low hammer. Figure 4.25 shows the hammering process. The perforated specimen is a central perforated aluminum plate with a thickness of 4 mm and the length direction is the rolling direction. The dimensions are the same as those of the cold expansion test piece, see Fig. 4.2.

The low hammer test method used is the drop hammer method. This method is to put the specimen on the lower part of the drop hammer test machine. The falling hammer strikes the specimen by free-fall motion. The hammer energy is determined by the height of the drop hammer and the quality of the hammer. Drop hammer test machine is the domestic Sans drop hammer test machine, and the maximum drop of the experimental machine height is 2 m. Sans drop hammer test machine work and the name of the parts are shown in Fig. 4.26. The hammer used is hemispherical, its radius is 50 mm, and the quality of the hammer can be adjusted according to the

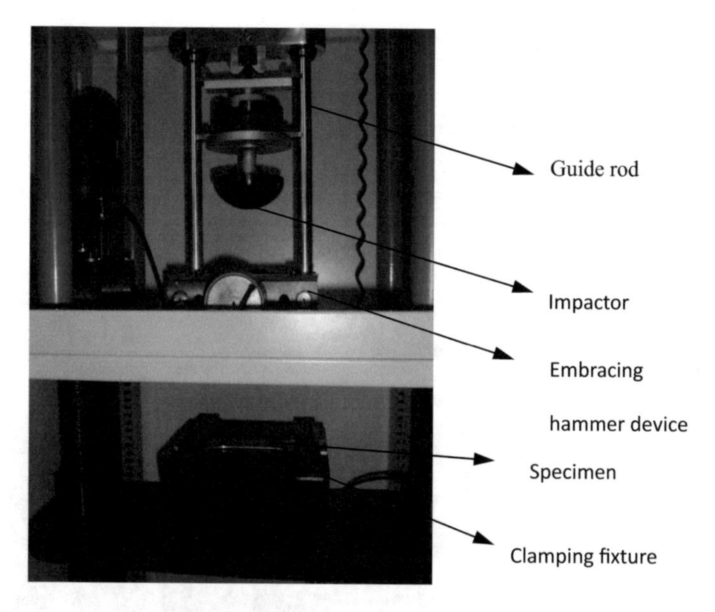

Fig. 4.26 Sans drop hammer test machine diagram

needs of the user through the replacement of internal weight. In this test, the mass of the hammer was 2.5 kg.

In this section, a total of 20 specimens were subjected to fatigue tests. According to the hammering energy or falling height, they were divided into four groups: A, B, C, and D, five per group. The drop height of group A is 0 m, that is, it does not do hammer test. B, C, and D are the hammering test. The drop height of group B is 1 m and the hammering energy is 24.5 J. Drop height of group C is 1.5 m, and hammering energy is 36.75 J. Drop height of group D is 2 m and hammering energy is 49 J.

The fatigue test of the hammered specimens was carried out on the homemade Letry resonant high-frequency fatigue testing machine (PLG-100). The loading method was stress control, using the sine wave amplitude fatigue load spectrum, the stress ratio of $R = 0.1$, the loading frequency of $f = 96$ Hz (The frequency determined by the test machine according to the specimen and loading method), and the maximum nominal stress is 120 MPa. The test was carried out at room temperature.

After fatigue test, the median fatigue life, standard deviation, and coefficient of variation of the hammers in each group are shown in Table 4.4.

The median fatigue life of each group is transformed into a graph; the median fatigue life of each group is shown in Fig. 4.27.

As can be seen from Table 4.4 and Fig. 4.27, the fatigue life of the openings after hammering increases under this size. As the drop hammer height or hammer energy increases, the fatigue life of the opening increases. When the hammer energy is 36.75 J (when the drop hammer height of 1.5 m), the hole fatigue life is the largest, about 5 times the life of non-hammer hole. When the hammering energy

Table 4.4 The median fatigue life, standard deviation, and coefficient of variation (2.5 kg)

Group	Median fatigue life/cycle	Standard deviation	Coefficient of variation
A	118,720	13,621.19672	0.1147
B	350,300	258,672.81457	0.7384
C	571,840	420,869.41918	0.736
D	475,800	256,850.41367	0.54

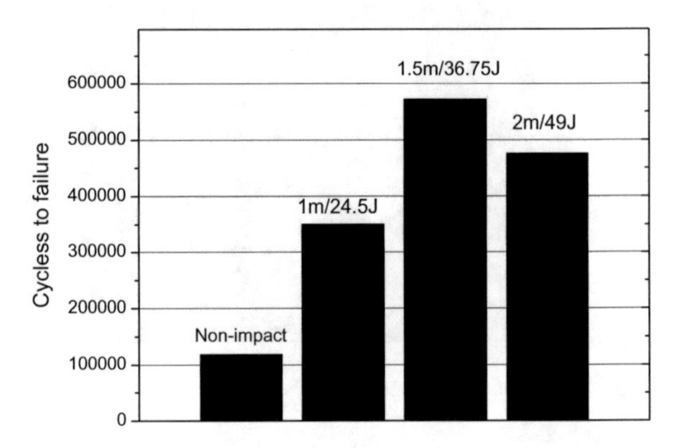

Fig. 4.27 Effect of hammer height on the fatigue life of the hole (drop weight 2.5 kg)

continues to increase, the fatigue life of the openings is slightly reduced. Obviously, not the greater the hammer energy is, the higher the fatigue life of the hole, but the existence of a best hammering energy. When the hammer energy is greater or less than this value, the fatigue life of the opening is reduced. Thus, according to the experimental results, 36.75 J is the optimum hammering energy for the size specimen. However, hammering to improve the life of the same time also led to increased dispersion of life, the best hammer energy of the hole life of the largest dispersion.

In addition, the results of the anti-fatigue test of the hammer method and the cold expansion method are shown in Tables 4.2 and 4.4. It can be seen that the maximum life after hammering is 571,840 cycles, and the maximum life after cold expansion is 645,645 cycles, cold expansion life is 1.1 times that of the hammer, and the dispersion of the fatigue life is larger after hammer. In general, the effect of the anti-fatigue effect of the hammering method is not as good as that of the cold expansion method. But in some special processing environment, the hammering method is also an alternative to the anti-fatigue enhancement of the orifice plate.

4.3.2 Finite Element Analysis of Hammering

The finite element model of different hammering heights is established. Since the structure and the load are symmetrical about one direction and two directions, only

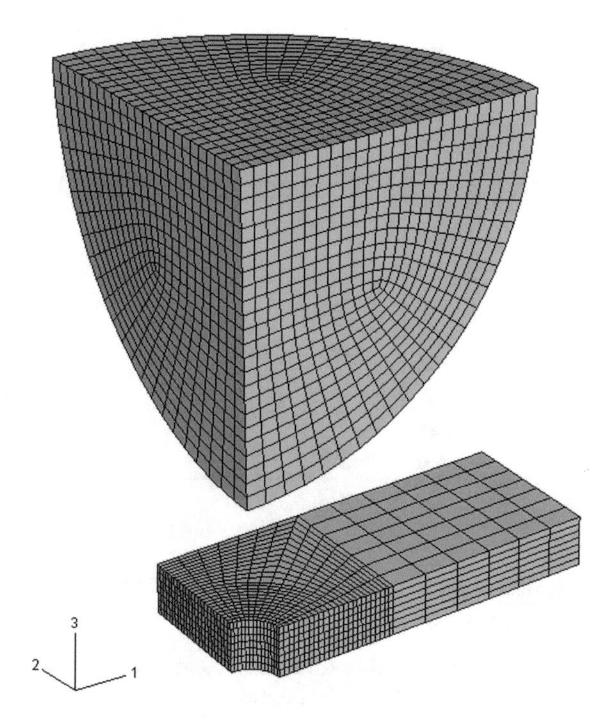

Fig. 4.28 Finite element model of hammer

1/4 of the original model is analyzed, as shown in Fig. 4.28. The initial speed of the drop hammer is 4,430 mm/s (corresponding to drop hammer height of 1 m), 5,422 mm/s (corresponding to falling hammer height of 1.5 m), and 6,261 mm/s (corresponding to drop hammer height of 2 m).

According to the calculation results of the tangential residual stress distribution of the perforated aluminum plate after hammering at different heights, the tangential residual stress distribution along the wall in the thickness direction is shown in Fig. 4.29.

It can be seen from Fig. 4.29 that the distribution of the hole walls in the thickness direction is very different at different hammering heights or hammering energies: at the falling height of 1 m, the tangential compressive residual stress initially decreased along thickness direction. By reaching a distance from the hammer surface of 0.5 mm, it started to increase. The maximum compressive tangential residual stress occurs at a thickness of 1.2 mm and a value of −370 MPa. The compressive tangential residual stress then begins to decrease in the thickness direction of the cell wall and decreases to a minimum of −160 MPa on the fixed surface. When the falling height is 1.5 and 2 m, the maximum compressive tangential residual stress occurs on the hammer surface, and the values are −392 and −401 MPa, respectively. And then the compressive tangential residual stress began to decrease, reduced to the thickness of the position of 1.2 mm and began to increase, increased to the thickness of the location of 1.8 mm, and began to decrease in the fixed surface to reduce the minimum value of −267 and −313 MPa.

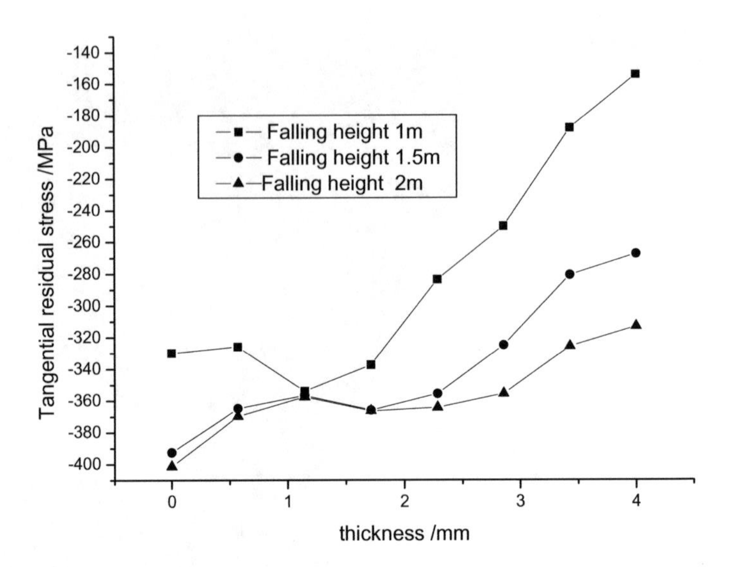

Fig. 4.29 The tangential stress distribution of the hole wall along the thickness direction

It can be seen that after the hammering, the distribution of the compressive residual stress in the thickness direction of the hole wall is not monotonic, but is fluctuating. The minimum compressive residual stress occurs on the fixed surface, and the maximum compressive residual stress varies according to the hammering energy, and its position is at the junction of the thickness of 1.2 mm and the hammer face and the hole wall.

At different hammering heights, the tangential residual stresses of the perforated plates are shown in Figs. 4.30, 4.31, and 4.32 on the hammer surface, the middle surface, and the fixed surface, respectively.

From Fig. 4.30, it can be seen that the maximum tangential compressive residual stress on the hammering plane occurs at about 0.5 mm from the orifice at different hammering heights. The higher the hammering height, the greater the maximum tangential residual stress on the hammer surface, the greater the compressive residual stress zone. The maximum tangential tensile residual stress occurs at 2–4 mm from the hole.

It can be seen from Fig. 4.31 that the maximum tangential compressive residual stress on the middle surface and the occurrence of the maximum tangential tensile residual stress are different at different hammering heights, when the hammering heights are 1 and 1.5 m. The maximum tangential compressive residual stress occurs at the orifice, and the maximum tangential tensile residual stress occurs at 2–4 mm from the hole. When the hammering height is 2 m, the maximum tangential compressive residual stress occurs at about 0.5 mm from the orifice, and the maximum tangential tensile residual stress occurs on the free edge of the model. The higher the hammering height, the greater the maximum tangential residual stress on the middle surface, and the larger the compressive residual stress zone.

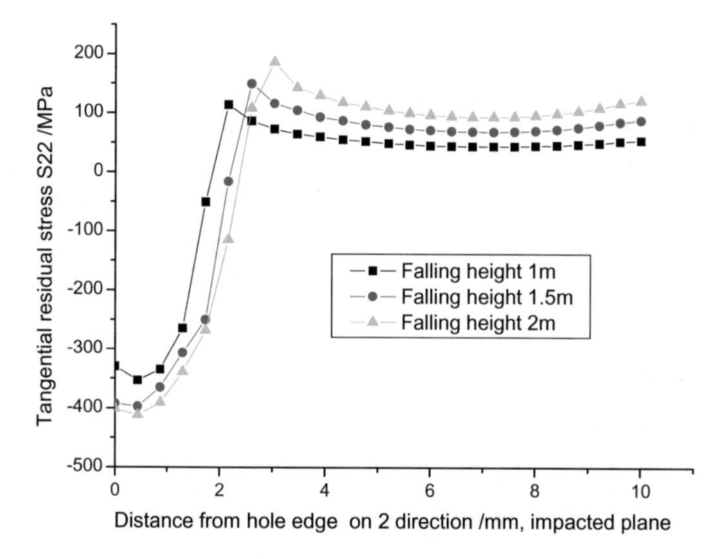

Fig. 4.30 Residual stress distribution on the hammer surface at different hammering heights

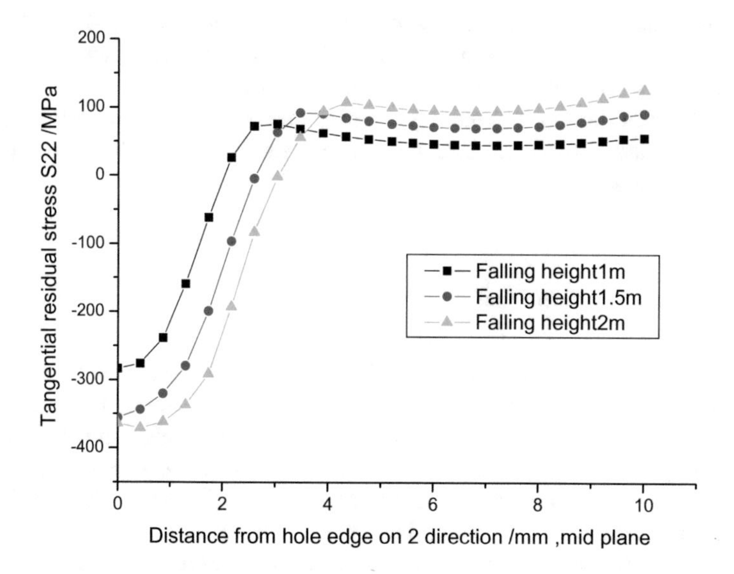

Fig. 4.31 Residual stress distribution of tangential tangent at different hammers

As can be seen from Fig. 4.32, the maximum tangential compressive residual stress on the fixed surface occurs at different hammers. The higher the hammering height, the larger the maximum tangential residual stress on the fixed surface, and the larger the compressive residual stress zone. The occurrence of the maximum tangential tensile residual stress is different, when the hammering height is 1 and

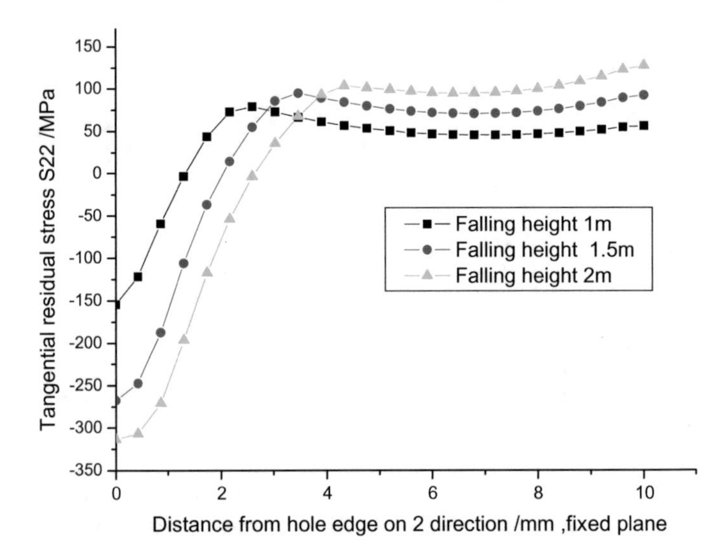

Fig. 4.32 Residual stress distribution of tangential tangent at different hammering velocities

1.5 m, the maximum tangential tensile residual stress occurs at 2–4 mm from the hole wall. When the hammering height is 2 m, the maximum tangential tensile residual stress occurs at the free edge of the model on.

It can be seen that the hammering height is different and the maximum residual tensile stress and the maximum compressive residual stress position on different layers are different.

4.3.3 Fracture Analysis

After the experiment, the fracture of all the specimens was analyzed by means of a microscope. It is found that after the hammering, the crack source forms are corner cracks, and all the corner cracks are initiated on the hammering surface of the open holes. The growth rate of the crack in the direction of the hammering plane is significantly larger than its growth rate in the thickness direction. The typical crack pattern of the hole after hammering is shown in Fig. 4.33.

From the finite element analysis of the previous section, it can be seen that the tangential residual compressive stress is the largest on the hammer surface, but the fatigue crack source always occurs on the hammer surface. The author believes that some damage in the hammer surface caused by hammering process is the main cause of cracks in the hammer surface. And after hammering, the fatigue life is improved by the hammering process caused by the introduction of compressive residual stress.

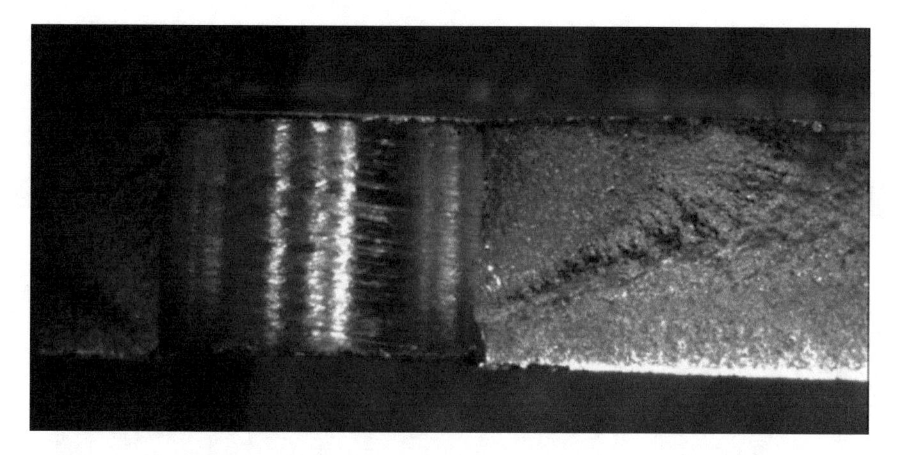

Fig. 4.33 Crack after typical hammering openings

Since the residual tangential compressive stress along surface direction rapidly reduced and eventually become residual tensile stress. The resistance of crack propagation reduced along surface direction and crack growth rate is speed-up. But residual tangential compressive stress always existed along thickness direction and then there was a resistance of crack growth and crack propagation retarded. This is the main reason why the crack expansion speed in the surface direction is significantly larger than its expansion speed in the thickness direction.

References

1. Wang, Min. 1999. *Principle and Technology of Anti-fatigue Manufacture*. Jiangsu: Jiangsu Science and Technology Press.
2. Karabin, M.E., and F. Barlat. 2007. Numerical and Experimental Study of the Cold Expansion Process in 7085 Plate Using a Modified Split Sleeve. *Journal of Materials Processing Technology* 189: 45–57.
3. Gopalakrishna, H.D., and H.N. Narasimha Murthy. 2010. Cold Expansion of Holes and Resulting Fatigue Life Enhancement and Residual Stresses in Al 2024 T3 Alloy—An Experimental Study. *Engineering Failure Analysis* 12: 361–368.
4. Yang, Hongyuan, and Wenting Liu. 2010. Experimental Study on Fatigue Life of Pore Extrusion. *Mechanical Strength* 32 (3): 446–450.
5. Chakherlou, T.N., and J. Vogwell. 2003. The effect of cold expansion on improving the fatigue life of fastener holes. *Engineering Failure Analysis* 10: 13–24.
6. Gong, Peng, and Linbing Zheng. 2011. Study on Extrusion Process Performance of 7B50-T7451 Aluminum Alloy Sheet. *Journal of Aeronautical Materials* 31 (4): 45–50.
7. Chakherlou, T.N., and M. Mirzajanzadeh. 2010. An Investigation About Interference Fit Effect on Improving Fatigue Life of a Holed Single Plate in Joints. *European Journal of Mechanics A/Solids* 29: 675–682.
8. Liu, Xiaolong, Yukui Gao, Yuntao Liu, and Dongfeng Chen. 2011. Three-Dimensional Finite Element Simulation and Experimental Study on Residual Stress Field Strengthened by Pore Extrusion. *Journal of Aeronautical Materials* 31 (2): 24–27.

9. Farhangdoost, Kh, and A. Hosseini. 2011. The Effect of Mandrel Speed Upon the Residual Stress Distribution around Cold Expanded Hole. *Procedia Engineering* 10: 2184–2189.
10. Yuan, Hongxuan. 2007. Manufacturing Technology of Connecting Brackets for Aircraft Structures. *Aeronautical Manufacturing Technology* 1: 96–99.
11. Wang, Haihong. 2007. Experimental Study on Compressive Fatigue Compression Process of Oil Hole Wall of Oil Hole. *Laboratory Research and Exploration* 26 (11): 248–250.

Chapter 5
Shot Peening Strengthening Technology

As the requirements of high speed, high maneuverability, lightweight, long lifetime and high reliability of aircraft become higher and higher, more and more integrated structures such as integral frame, integrated panel and integrated wing covering, are widely used in aircraft design. To improve the fatigue life of integrated structure, shot peening are applied on structure after machining [1, 2]. In this chapter, the influence of shot peening on fatigue life and residual stress field of components are studied by using finite element method and related experimental techniques.

5.1 Mechanism of Shot Peening and Status of Art

Surface shot peening technology emerged in the 20s of last century, which has been widely used in the machinery and aviation industry in the 60s. Shot peening is a method of cold working on the high-speed impact of metal parts surface with small hard shot flow [3], which introduced the compressive residual stress, work hardening and changed the surface roughness of the workpiece. In addition, stress induced phase transformation and deformation induced crystal texture transformation occur near the surface during shot peening. So, the fatigue crack is not easy to initiate on the surface of the part and mainly occurs in the secondary surface layer. However, the secondary surface layer is constrained by the surrounding crystal grain; a greater critical stress is required for the crystalline slip and crack initiation, so the fatigue life of the parts is improved [4–6]. As a simple and effective method of surface strengthening, shot peening can significantly improve the strength and fatigue fracture resistance, the life of stress corrosion, and hydrogen cracking fracture resistance, and has the advantages of simple, wide application, and low energy consumption.

In the process of shot peening, high-speed shot continuously impacts on the surface of the workpiece, which is equivalent to the countless mini-hammering metal surface. The diameter of the crater formed by the impact of shot is smaller

J. Liu et al., *Long-Life Design and Test Technology of Typical Aircraft Structures*,
https://doi.org/10.1007/978-981-10-8399-0_5

than the diameter of shot, as well as the depth. On the whole surface, the craters are relatively flat. The microstructure has also been changed, such as the sharp increase of dislocation density and the emergence of subgrain boundaries and grain refinement. If the dislocation gradually arranged regularly, the structure will be changed again to stable state.

The shots that mainly used in shot peening are the steel shots, ceramic shots, and glass shots. There are different residual compressive stress fields when the shot impacts the target, and the depth of the compressive stress layer is also different. The number of shots used in shot peening is huge, and the number of impact times per unit area is very striking. In this way, everywhere of workpiece can get uniform shot opportunity. The parameters of the shot peening strengthening effect are as follows: the size, velocity, hardness, flow of the shot, the spraying time, and the coverage. Any change in any of these parameters will affect the strengthening effect of the workpiece surface.

The main advantages of shot peening technology are as follows:

1. improve component life;
2. reduce cost;
3. avoid notch sensitivity;
4. reduce the requirements of fine surface finish.

At present, scholars at inland and abroad have done a lot of research on the principle of shot peening, residual stress field, and finite element modeling [7–15]. Dalaei et al. studied the effect of shot peening on the fatigue life of pearlitic microalloyed steels and concluded that the shot peening process introduced a stable residual compressive stress and studied the effects of shot peening on the constant load fatigue [7, 8]. Gangarj et al. believed that shot peening improves the fretting fatigue strength of the critical parts, and used finite element method to simulate normal stress, shear stress, body stress, and displacement amplitude [9]. Gao Yukui studied the relaxation law of residual stress on the surface of ultrahigh strength steel shot peening [10]. Liu et al. studied the effect of shot peening on the high cycle fatigue properties of Mg-10Gd-3Y alloy under different conditions [11]. The effects of strain hardening rate, shot peening coverage, and initial residual tensile stress on the residual stress distribution of 304 stainless steel target were studied by Xiang et al. [12]. The effect of shot peening on the fatigue properties of composites was studied by Chen Bo et al. Zhang Weihong found that the residual stress field of the large shot is deep, and the residual stress can be increased by increasing the shot peening coverage by 2024 aluminum alloy shot peening experiment [14]. According to the numerical simulation of shot peening of 7075 aluminum alloy by Zhang Hongwei, it is found that the increase of shot peening coverage will significantly change the residual stress field and make the distribution of residual stress field more uniform [15].

5.2 Effect of Shot Materials on Fatigue Performance

On shot material, in the 60s of last century, the aviation industry mainly used glass shot, then cast iron and cast steel shot. In this section, the effects of different shot material including glass shot and cast steel shots, etc., on the fatigue life of the components were studied by means of experiments.

5.2.1 *Fatigue Test*

Shot peening fatigue test piece was shown in Fig. 5.1. The geometric dimensions are as follows: $L_1 = 250$ mm, $L_2 = 45$ mm, $L_3 = 18$ mm, $R = 110$ mm, and the plate thickness is 8 mm. The test pieces were divided into three groups named A, B, and C. Each group contains eight specimens. All the test pieces experienced the same machinery processing technology. A group of test pieces were shot peened by cast steel shot, B group of test pieces were shot peened by glass shot, and C group of test pieces were not shot peened. The shot peening coverage of all test pieces was 100%.

Fatigue test was performed on INSTRON 8801 hydraulic servo testing machine, as shown in Fig. 5.2. The loading frequency is 10 Hz, the maximum load is 250 MPa, the stress ratio is $R = 0.1$.

5.2.2 *Comparative Analysis of Fatigue Test Data*

The statistical results of fatigue test data are shown in Table 5.1.

After logarithm of treatment on the fatigue life of the three groups in Table 5.1, *t*-test was used to evaluate the effect of surface peening between any two group.

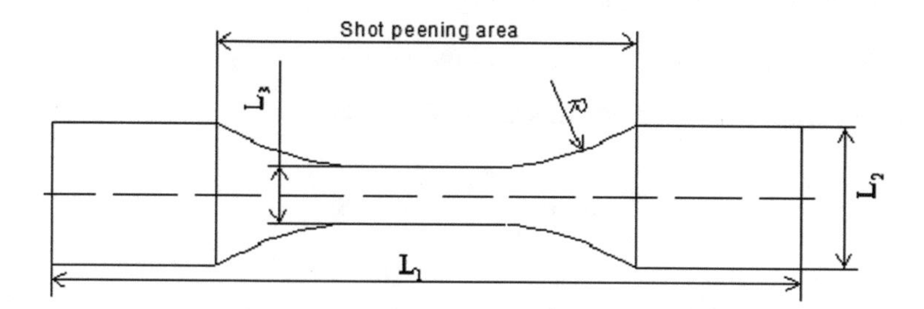

Fig. 5.1 Shot peening strengthening test piece

Fig. 5.2 Shot peening fatigue test device

Table 5.1 Fatigue test life results

Specimen type	Cast steel shot peening (group A)	Glass shot peening (group B)	Non-shot blasting parts (group C)
Fatigue life/cycle	162,398	176,264	151,882
	121,880	118,750	109,301
	156,727	97,130	80,082
	234,456	146,431	71,902
	272,248	96,276	81,288
	220,240	127,410	103,707
	264,410	84,277	87,105
	179,097	224,595	167,540

Comparison between A group and C group:

(a) Calculate mean $\bar{x}$ and variance s^2 of the sample

For the specimens of group A, sample size $n_1 = 8$

$$\overline{x_1} = \frac{1}{n_1}\sum_{i=1}^{n_1} x_{1i} = 5.289 \tag{5.1}$$

$$s_1^2 = \frac{1}{n_1 - 1}\sum_{i=1}^{n_1}(x_{1i} - \overline{x_1})^2 = \frac{1}{n_1 - 1}\left(\sum x_{1i}^2 - n\overline{x_1}^2\right) = 0.015 \tag{5.2}$$

For the specimens of group C, sample size $n_3 = 8$

$$\overline{x_3} = \frac{1}{n_3}\sum_{i=1}^{n_3} x_{3i} = 5.009 \tag{5.3}$$

$$s_3^2 = \frac{1}{n_3 - 1}\sum_{i=1}^{n_3} (x_{3i} - \overline{x_3})^2 = \frac{1}{n_3 - 1}\left(\sum x_{3i}^2 - n\overline{x_3}^2\right) = 0.018 \tag{5.4}$$

(b) F-test

Before applying the t-test, the F-test must be performed. When the F value is calculated, the larger variance 0.018 is taken as a molecule and the smaller variance is 0.015 as the denominator, getting the result:

$$F = \frac{0.018}{0.015} = 1.187 \tag{5.5}$$

The molecular degrees of freedom correspond to the larger variance, the value is 8 – 1 = 7 and the denominator freedom is 8 – 1 = 7. Take a significant degree of $\alpha = 5\%$, then check the table $F_\alpha = 4.99$, by $F < F_\alpha$. So, the two population root-mean-square equals $\sigma_1 = \sigma_3$.

(c) t-test

Calculate t value:

$$t = \frac{(\bar{x}_1 - \bar{x}_3)}{\sqrt{s_1^2 + s_3^2}}\sqrt{n} = 4.37 \tag{5.6}$$

Its degrees of freedom is

$$v = n_1 + n_3 - 2 = 8 + 8 - 2 = 14$$

Take $\alpha = 5\%$, and check the table by $t_\alpha = 2.145$, by $t > t_\alpha$. As a result, in terms of fatigue life, it can be concluded that the fatigue life of the two specimens are significantly different, by $\bar{x}_1 > \bar{x}_3$, we can see that the fatigue life of cast steel shot peening parts is higher.

(d) interval estimation

In order to compare the fatigue life of cast steel shot peening parts than that of non-shot peening parts, the confidence values is chosen as $\gamma = 95\%$. There is still a check follows $t_\gamma = 2.145$, because of $\bar{x}_1 > \bar{x}_3$, there is an interval estimation:

$$(\bar{x}_1 - \bar{x}_3) - t_\gamma s_{13}\sqrt{\frac{1}{n_1} + \frac{1}{n_3}} < \mu_1 - \mu_3 < (\bar{x}_1 - \bar{x}_3) + t_\gamma s_{13}\sqrt{\frac{1}{n_1} + \frac{1}{n_3}} \tag{5.7}$$

The upper and lower confidence limits, respectively, are

$$(\bar{x}_1 - \bar{x}_3) + t_\gamma s_{13}\sqrt{\frac{1}{n_1} + \frac{1}{n_3}} = (5.29 - 5.01) + 2.145 \times 0.129 \times 0.5 = 0.41892 \tag{5.8}$$

$$(\bar{x}_1 - \bar{x}_3) - t_\gamma s_{13}\sqrt{\frac{1}{n_1} + \frac{1}{n_3}} = (5.29 - 5.01) - 2.145 \times 0.129 \times 0.5 = 0.14222 \tag{5.9}$$

95% confidence interval estimation

$$0.14222 < \mu_1 - \mu_3 < 0.41892 \tag{5.10}$$

Write in the form (5.7)

$$0.14222 < \lg\frac{[N_{50}]_1}{[N_{50}]_3} < 0.41892 \tag{5.11}$$

In the form of the antilog taking the value of equation above all, then get.

$$1.387458 < \frac{[N_{50}]_1}{[N_{50}]_3} < 2.623735 \tag{5.12}$$

According to the above equation, the median fatigue life of cast steel shot peening is 1.39–2.62 times of the fatigue life of the shot peening under the confidence of 95%.

Repeat the above steps, we can draw a conclusion: under the confidence level of 95%, blasting glass piece's median fatigue life is 0.88–1.76 times to non-shot peened specimens median fatigue life; cast steel shot peened specimens median fatigue life is 1.09–2.13 times to blasting glass piece's median fatigue life. Therefore, the shot peening can improve the fatigue life of the component, and the strengthening effect of the cast steel pill is better than that of the glass pill.

5.2.3 Fatigue Fracture Analysis

The residual compressive stress field and the depth of compressive stress layer formed by different shot material have different effects on the fatigue fracture morphology. The fracture surface of all the test pieces was analyzed by using XYH-06 stereo microscope. The typical fracture of non-shot peening, glass, and cast steel shot peening specimens was given in Fig. 5.3. As can be seen from Fig. 5.3, the fatigue cracks of the non-shot peened parts are formed on the

Fig. 5.3 Fracture morphology of shot peening under different shot peening mediums.

(a) Typical fracture surface of non-shot peening strengthening specimen

(b) Typical fracture surface of glass bead peening strengthening

(c) Typical fracture surface of cast steel peening strengthening

nucleation surface of the specimen in the form of corner cracks, and the fatigue crack propagation area is formed by the outward expansion of the sector, then formed crack propagation zone. And the fatigue cracks of the shot peening parts on subsurface nucleation expand outward. Due to the effect of residual compressive stress, in depth direction, the crack growth length of the shot peening specimens is smaller than that of the non-shot peened specimens. In addition, the fatigue fracture surface of the shot peening parts is more glossy than that of the non-shot peened parts. This is due to the slower crack growth rate and the longer time of friction between the crack opening surfaces, which leads to the appearance of shell-like luster.

5.2.4 *Numerical Simulation of Shot Peening with Different Shot Materials*

In this section, 3D finite element simulation and analysis of multiple shot peening processes were carried out by ABAQUS. The residual compressive stress field and the depth of compressive stress layer after shot peening were analyzed with respect to the cast steel shot and glass shots, which provided the theoretical basis and analysis method for the analysis of the anti-fatigue effect of different shot materials.

5.2.4.1 Material Parameters

Aluminum alloy 2024 was used for the specimens. The elastic modulus of is $E = 72$ GPa, Poisson's ratio is $v = 0.3$, and density is $\rho = 2600\ \text{kg/m}^3$. It is assumed that the yield of the target obeys the Mises yield criterion, the yield strength is $\sigma_y = 500$ MPa, the plastic deformation stage is assumed to be linear strain hardening, and the strain hardening rate is $H = 500$ MPa [12]. The basic mechanical properties of the shot are shown in Table 5.2. The shot radius is $r = 0.8$ mm, and the target is 1/4 cylinders. In order to eliminate the influence of boundary conditions, the radius and height of the target are $10r$.

Table 5.2 Mechanical properties of cast steel shots and glass shots [16]

Shot type	Density ρ (kg m^{-3})	Elastic modulus E/GPa	Poisson's ratio μ
Cast steel shots	5500	210	0.3
Glass shots	2800	7.919	0.3

5.2.4.2 Model Building

In order to simplify the calculation, only the 1/4 symmetry model is considered. The finite element model is shown in Fig. 5.4. *X* direction constraint was applied on *Y-Z* symmetry plane; *Y* direction constraint was applied on *X-Z* symmetry and *Z* direction constraint was applied on bottom surface. For shot, only lower half part which contact with target was used and shot only move in *Z* direction. The speed of shot $v = 100$ m/s.

Dynamic/Explicit solver in ABAQUS was employed. The normal direction of the shot and the target is defined as the hard contact, and the tangential direction is defined as the frictionless contact mode. In order to obtain the stable residual stress field and save the computational cost, the analysis time is set to 4×10^{-5} s. It is proved by calculation that when the calculation is finished, the collision process is over, the shot is bounced, and the residual stress field is stable (which does not change with time).

5.2.4.3 Simulation Results

The residual stress distribution after shot peening is shown in Fig. 5.5. The residual stress of the surface of the target is caused by the mismatch of elastic–plastic deformation. According to the results of finite element simulation, the distribution of the residual stress and the depth of the compressive stress layer are different too. Figure 5.6 analyzed the variation of the residual stress along the *Z* direction on the central axis of the target. It can be seen that the size and location of the residual

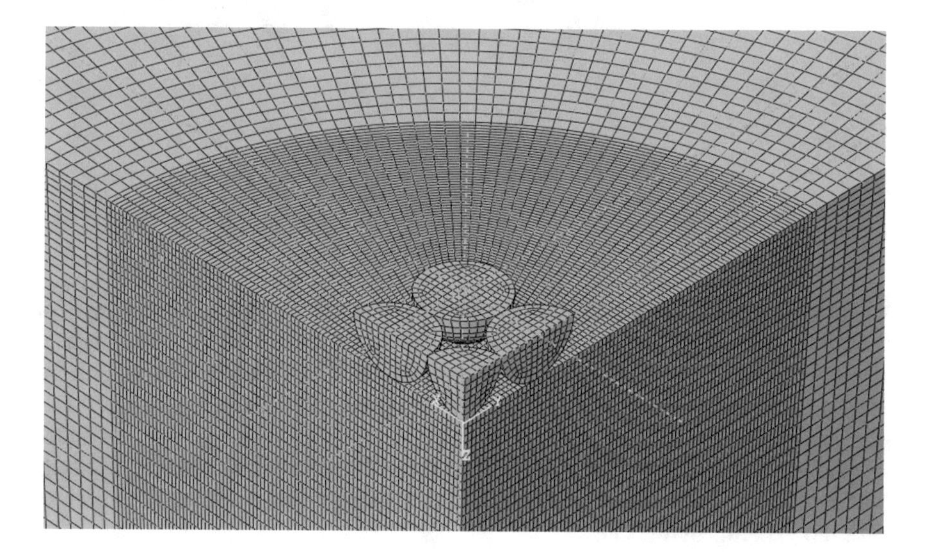

Fig. 5.4 Shot peening finite element model

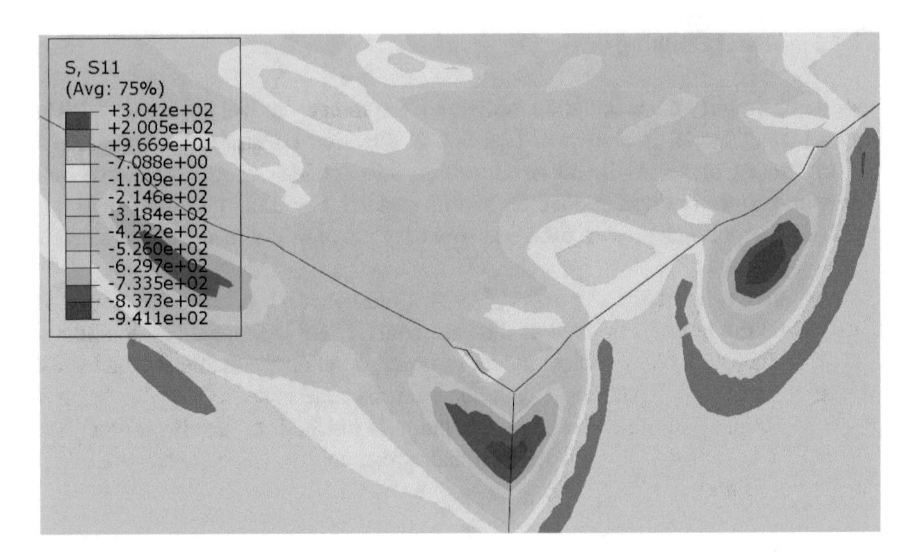

Fig. 5.5 Residual stress field distribution in target shot peening area

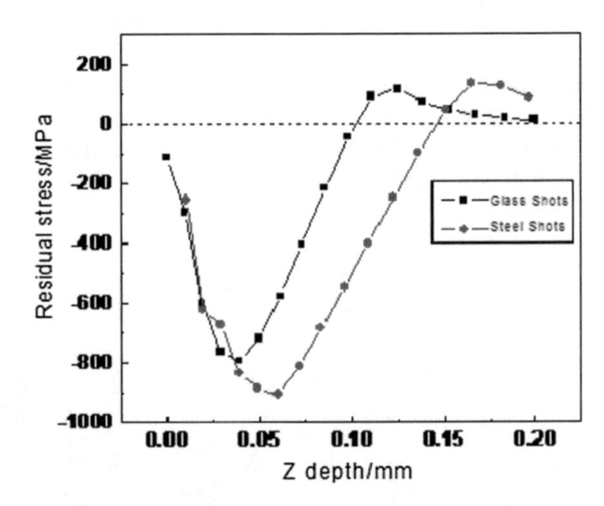

Fig. 5.6 The residual stress distribution of the central pellet introduction after shot peening

compressive stress and the depth of the residual compressive stress layer vary with the type of shot. Steel shot after shot peening of the maximum compressive residual stress (904 MPa), the maximum residual stress position (about 0.6 mm) and the depth of the stress layer (about 0.15 mm) were greater than the maximum residual compressive stress of glass shots after shot peening (794 MPa), the maximum residual stress position (about 0.4 mm) and residual compressive stress layer depth (about 0.10 mm). The effect of cast steel shot peening is better, which is consistent with the experimental results.

5.3 Effect of Surface Roughness on Residual Stress Field of Shot Peening

Most of the researches on shot peening do not take the influence of the roughness of the target into account. In fact, the surface of the target may not be ideal, but it has to do with the processing and machining accuracy. The study of the influence of the surface roughness of the target on the residual stress field of shot peening is of great significance to improve the processing technology and optimize the shot peening parameters.

Generally speaking, there are four main characteristics of the residual stress field of shot peening [16]: the surface residual stress of σ_{srs}, the maximum compressive residual stress σ_{mrs}, the maximum compressive residual stress depth of σ_m, and compressive residual stress field depth of σ_0. This section focuses on the qualitative research on target surface roughness R_a of shot peening residual stress field, while the surface roughness and smooth surface under two conditions of influence of jet velocity and the shot diameter on residual stress field were compared, in order to find out the influence of the surface roughness of the target on the residual stress field of shot peening.

5.3.1 Finite Element Model

In this paper, finite element analysis is carried out by ABAQUS, and the rough surface is simplified to cosine surface. The definition of the parameters used in this paper is shown in Fig. 5.7, h is the cosine half-amplitude, s is 1/4 wavelength, and R is the radius of the shot. In this paper, profile arithmetic average error R_a is the basic parameters used to measure the surface roughness:

$$R_a = \frac{1}{l}\int_0^l |y(x)|\mathrm{d}x \tag{5.13}$$

In Eq. (5.13), $y(x)$—the height of the surface profile based on the centerline; l—the length of the sample. The R_a parameter contains a lot of information about the micro-roughness, which can be used to characterize the practical property of surface roughness. Keep cosine wavelength s = 0.125 mm unchanged, by changing the h value can get different surface roughnesses.

In order to simplify the calculation, a two-dimensional axisymmetric finite element model is used. As shown in Fig. 5.8, the shot is a rigid body with a radius of R, with a density of 5500 kg/m^3. The size of the target is chosen as $10R \times 10R$ to ensure that the boundary stress and strain are small enough. The mesh near the impact location is refined and exists stress concentration in the peak and notch, so the mesh is further refined. The four node axisymmetric linear

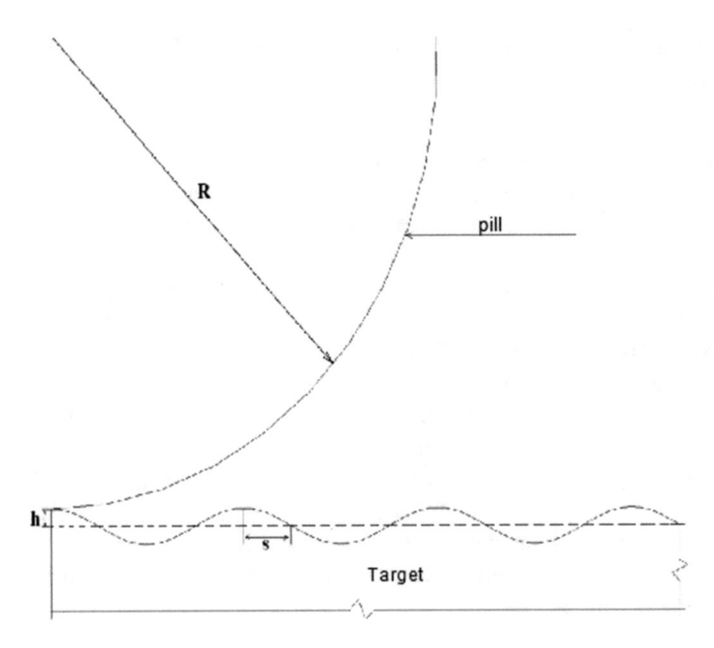

Fig. 5.7 The schematic diagram of parameters used in this paper

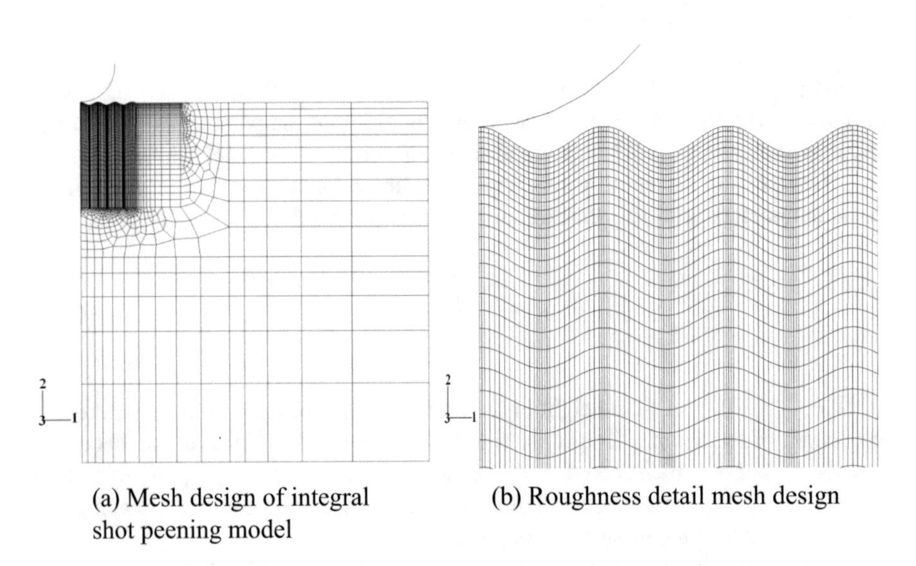

Fig. 5.8 Mesh design of shot peening finite element model considering the surface roughness of the target

reduced integral element (CAX4R) is used, and each model is divided into about 6000 units. The normal direction of the shot and the target is defined as the hard contact, and the tangential direction is defined as the frictionless contact mode. On the symmetry plane and the bottom surface, the normal direction is restrained, and the shot is allowed to move freely in the direction 2 (reference coordinate system).

5.3.2 Residual Stress Distribution

Because of the existence of surface roughness, the distribution of residual stress in surface of shot peening is different from that when the surface is smooth. When shot diameter R = 2 mm, shot peening speed v = 200 m/s, Fig. 5.9 shows the distribution of the transverse residual stress (S_{11}) in the two cases of smooth surface and rough surface (R_a = 12.7 μm).

As can be seen from Fig. 5.9a, when the target surface is smooth, there is a local residual compressive stress zone near the crater. Along the axis of symmetry, the residual compressive stress increases gradually, and the contour of the residual compressive stress zone is more smooth and full, and the residual compressive stress decreases gradually with the increase of depth, until it is transformed into residual tensile stress.

When the target surface is rough (Fig. 5.9b), there is a significant local residual compressive stress on the hollow site near the crater. The residual tensile stress is produced obviously at the center of the impact, and the residual compressive stress area becomes thinner and thinner compared with the smooth surface.

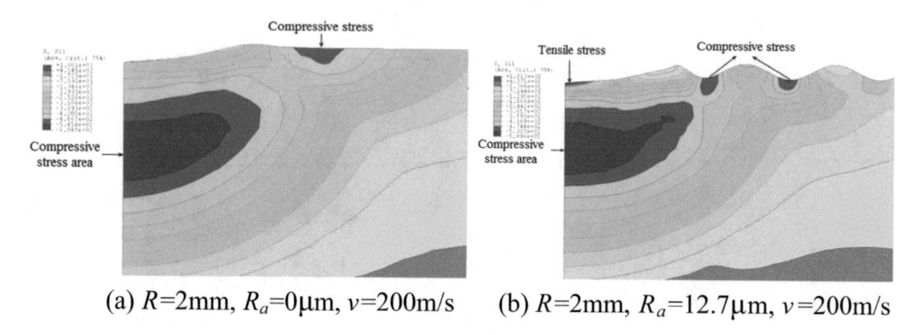

Fig. 5.9 Residual stress distribution characteristics of shot peening under different surface roughnesses

5.3.3 Effect of Surface Roughness on Residual Stress of Shot Peening

Considering the transverse residual stress distribution of the four different surface roughnesses (R_a = 0, 6.4, 12.7 and 25.5 μm). Figure 5.10a gives the variation of the transverse residual stress with depth z at four different roughness levels when the shot velocity is v = 100 m/s. As can be seen from Fig. 5.10a, the surface residual stress is compressive stress when the target surface is smooth. With the increase of surface roughness, the surface residual compressive stress decreases, until it is transformed into surface tensile stress and increases with the increase of roughness. In the selection range of R_a = 0–6.4 μm, the maximum residual compressive stress increases with the increase of roughness, and in the range of R_a = 6.4–25.5 μm, it decreases with increasing roughness. The maximum residual compressive stress depth and residual compressive stress field depth decrease with the increase of roughness.

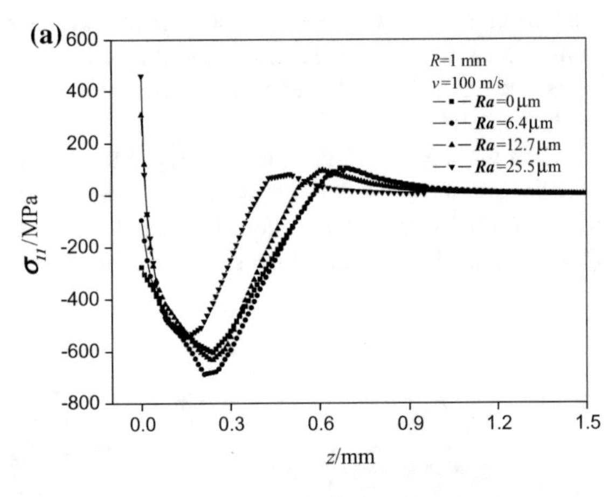

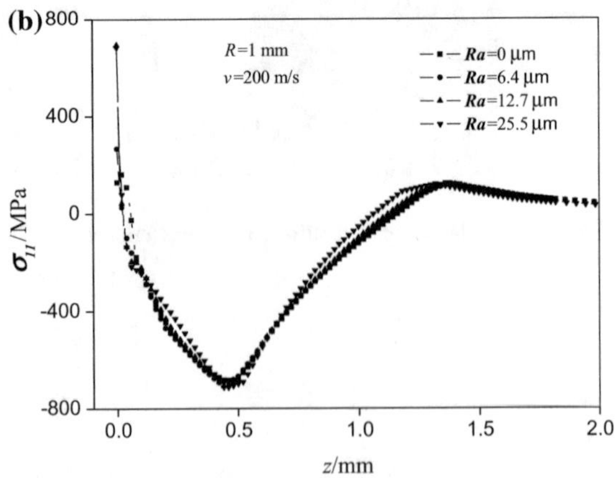

Fig. 5.10 Effect of surface roughness on residual stress of shot peening

Figure 5.10b gives the variation of the transverse residual stress with depth under the four surface roughness of $v = 200$ m/s. It can be seen from Fig. 5.10b that the influence of surface roughness on the residual stress and its distribution decreases with the increase of shot peening velocity. Compared with Fig. 5.10a, b, it is found that the increase of shot velocity will promote the transformation of surface residual stress from compressive stress to tensile stress.

Based on the above analysis, it can be seen that the increase of the surface roughness will lead to the decrease of residual compressive stress, and even bring the residual tensile stress on the impact surface, which leads to the thinning of the residual compressive stress zone. Therefore, it is necessary to improve the machining accuracy and make the surface of the target as smooth as possible in order to obtain the ideal residual stress distribution.

5.3.4 *Influence of Shot Size on Residual Stress Considering Surface Roughness*

The effects of three different shot sizes ($R = 1, 2, 3$ mm) on residual stress are considered. Figure 5.10a gives the variation of the transverse residual stress with depth z in the case of the shot velocity $v = 100$ m/s when the surface is smooth. It can be seen in Fig. 5.11a that when the surface is smooth, the shot size has no effect on the surface residual stress and the maximum residual stress, while the maximum compressive residual stress depth and residual compressive stress field depth increase with the increase of the shot radius.

Figure 5.11b gives the variation of the transverse residual stress with depth under different shot sizes when the shot velocity $v = 100$ m/s, $R_a = 12.7$ μm. As can be seen from Fig. 5.11b, when the surface roughness value is large, the surface residual compressive stress will be changed into tensile stress. When $R = 2$ mm, surface residual stress and the residual stress are all minimum. Checking the deformation after shot peening (see Fig. 5.12), it is found that when $R = 2$ mm, the deformation of the target just reaches the bulge of the surface. When $R = 1$ and 3 mm, the target deformation just reached the concave surface. The depth of the maximum residual compressive stress and the depth of the residual compressive stress field all increase with the increase of the shot radius.

Figure 5.11c gives the variation curve of transverse residual stress with depth when the shot velocity is $v = 200$ m/s. Compared with Fig. 5.11a–c, when the shot peening speed increases, the impact of these trends weakened.

Fig. 5.11 Effect of shot radius on residual stress

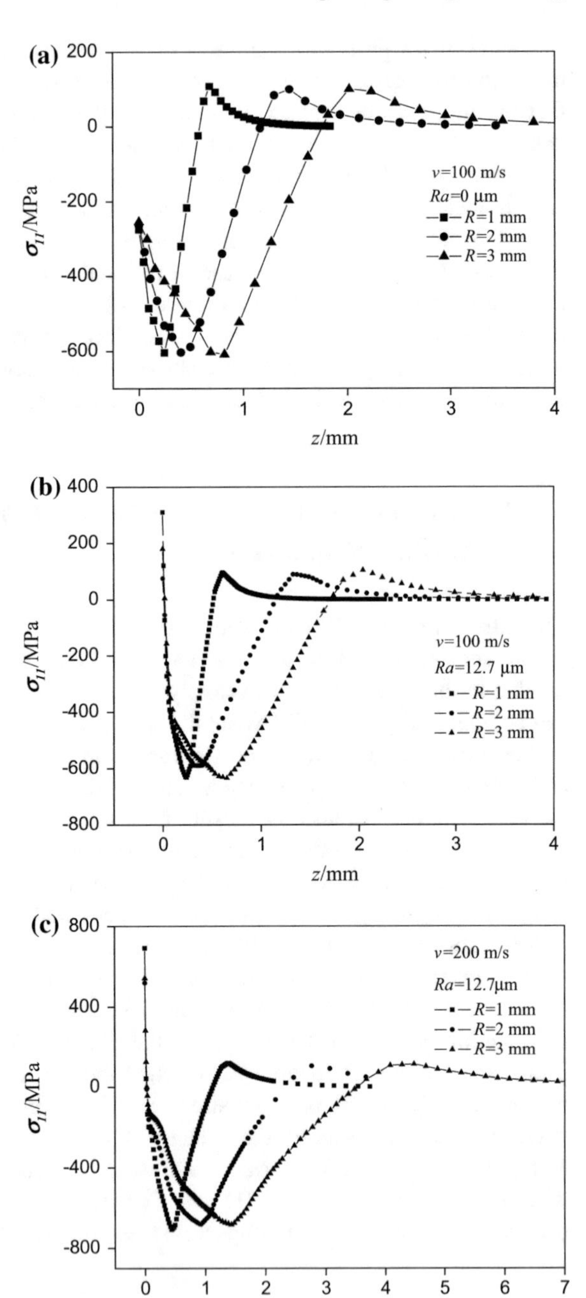

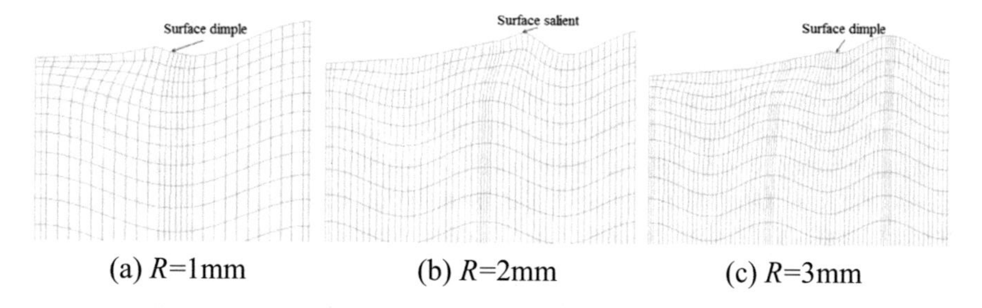

Fig. 5.12 Deformation diagram of shot with different sizes

5.3.5 The Effect of Shot Peening Velocity on Residual Stress Considering Surface Roughness

According to the analysis of the previous section, the effect of surface roughness on the residual stress of shot peening is decreased when the shot peening velocity increases. Therefore, it is necessary to study the two stages of low-speed and high-speed shot peening. In order to explain the influence of shot peening velocity on residual stress field clearly, the variation of the four residual stress field characteristics with the shot peening velocity in the smooth and rough surface under the two conditions is studied: the surface residual stress, the maximum residual stress, the maximum residual stress depth, and residual stress field depth, see Fig. 5.13c, d.

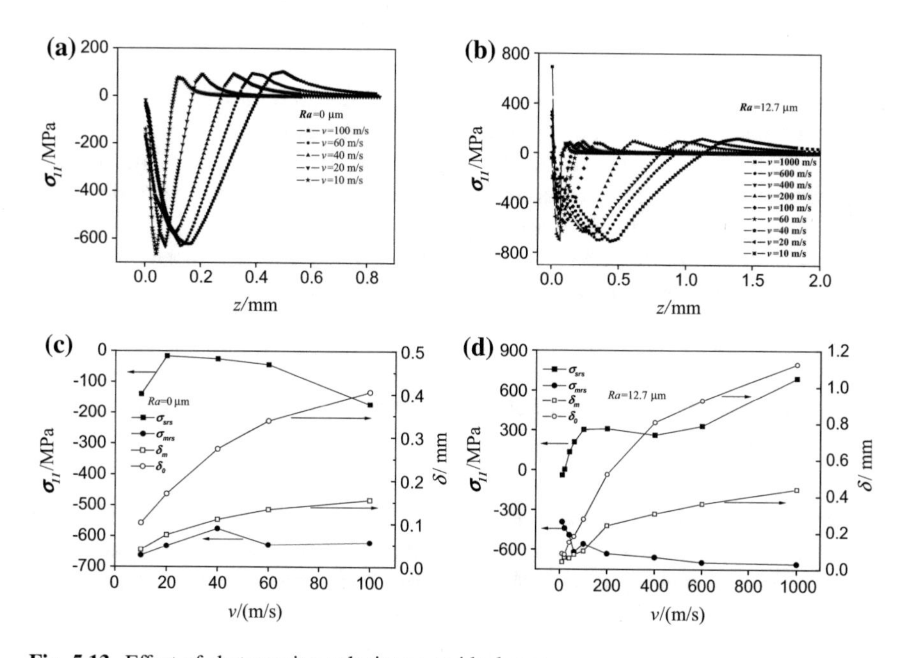

Fig. 5.13 Effect of shot peening velocity on residual stress

Figure 5.13a shows the effect of shot peening velocity on residual stress at the low-velocity impact stage (10–100 m/s). Refer to Fig. 5.13c, it can be seen that the surface residual compressive stress decreases with the increase of velocity in the range of 10–20 m/s, and then increases with the increase of velocity. The maximum residual compressive stress decreases with the increase of velocity in the range of 10–40 m/s and then increases with the increase of velocity. In the low-velocity impact stage, the larger velocity may not be able to obtain the large residual compressive stress. The maximum residual compressive stress depth and residual compressive stress field depth increase with the increase of shot peening velocity.

Figure 5.13b gives the effect of shot peening speed on residual stress when the target surface is rough (R_a = 12.7 μm). Refer to Fig. 5.13d, it can be seen that when the surface of the target is rough, the residual compressive stress of the surface changes to the residual tensile stress and increases with the increase of the shot peening velocity. The maximum residual compressive stress increases with the increase of shot peening velocity. The maximum residual stress depth and residual compressive stress field depth increase with the increase of shot peening velocity. When the shot peening velocity increases further, the influence trend tends to be stable, that is, the surface residual tensile stress, the maximum residual compressive stress, and its depth, and the depth of residual compressive stress field increases with the increase of shot peening velocity.

5.4 Effect of Mechanical Properties of Target Materials on Shot Peening Energy Conversion

Domestic and foreign scholars have done a lot of research on residual stress field and its influencing factors, and have carried on the experimental verification to the shot peening strengthening effect [17–20], but ignored the effect of mechanical properties of target on residual compressive stress. What is more, the problem of energy conversion during shot peening is rarely reported in [20]. The energy conversion in shot peening directly reflects the efficiency of shot peening, so it is necessary to study the shot peening process from the angle of energy.

5.4.1 Conversion Between Kinetic Energy and Deformation Energy

The symbols used in this paper and their meanings are shown in Table 5.3. The energy conversion during shot peening can be divided into two stages. First of all, the initial kinetic energy of the shot is transformed into the deformation energy of the target at the time of shot penetration. Second, in the rebound stage, the elastic deformation of the target material is restored, and the part of the kinetic energy is

Table 5.3 Symbols used for energy conversion in this section and their implications

E	Young's modulus	ΔE	Loss kinetic energy of system
υ	Poisson's ratio	σ_{11}	Transverse residual stress
σ_y	Yield strength	d	Distance from surface
H	Strain hardening rate	R	Shot radius
D_d	Crater diameter	C	Coverage rate
σ_{mrs}	Maximum residual compressive stress	σ_{srs}	Surface residual stress
Z_0	Zero residual stress depth	Z_m	Maximum residual compressive stress depth
ρ	density	E_p	Plastic strain energy
v_i	Initial incidence velocity	E_e	Elastic strain energy
v_r	Rebound velocity	E_{Total}	Total energy of system
e_r	Velocity return rates	E_d	Dissipated energy
α	Incident angle	ALLKE	System kinetic energy
t	Time	ALLIE	System strain energy
K	Energy conversion rate	m	Shot mass

returned to the shot. In the case of pure elastic contact, there is no energy loss in the impact process, and the incident kinetic energy is equal to the rebound kinetic energy. If it is elastic–plastic contact, only part of the kinetic energy returned after the impact. In addition, some kinetic energy is transformed into vibration and surface heat.

As shown in Fig. 5.14, the shot with the mass of m in initial velocity v_i and incident angle shot in and exit angle α' and speed v_r rebound. Define rate of return:

$$e_r = \frac{v_r \sin \alpha'}{v_i \sin \alpha} \tag{5.14}$$

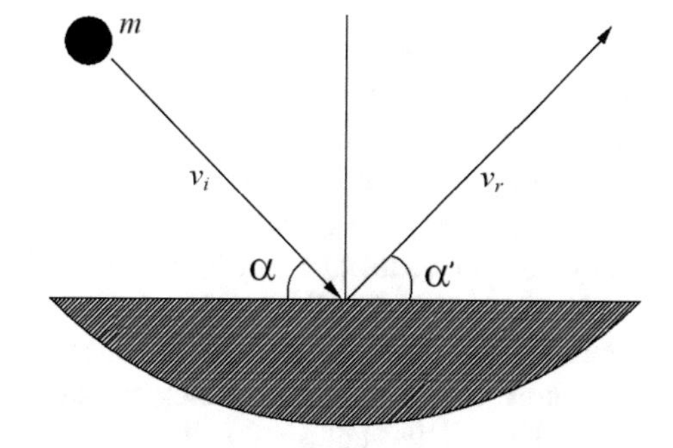

Fig. 5.14 Schematic diagram of shot peening process

It is obvious that there is $e_r = 1$ for the pure elastic impact process, but there is $0 < e_r < 1$ for the elastic–plastic impact process. In this paper, we consider that the shot is an ideal rigid body, that is, the plastic strain energy is not stored in the impact process. The kinetic energy loss of the shot is

$$\Delta E = \frac{1}{2} m v_i^2 (1 - e_r^2) \tag{5.15}$$

The kinetic energy of the shot loss is converted into two parts: (1) the plastic deformation energy of the target is E_p; (2) the energy dissipation is E_d (for example, the vibration and the surface heat, etc.):

$$\Delta E = E_p + E_d \tag{5.16}$$

In addition, without considering the influence of internal force and friction (the friction coefficient is 0), that is, the energy dissipation is zero, so the energy loss of the shot is all converted into the plastic deformation energy of the target:

$$\Delta E = E_p \tag{5.17}$$

Energy conversion coefficient K is defined as

$$K = \frac{\Delta E}{E_{\text{Total}}} = \frac{E_p}{E_{\text{Total}}} = \frac{\frac{1}{2} m v_i^2 (1 - e_r^2)}{\frac{1}{2} m v_i^2} = 1 - e_r^2 \tag{5.18}$$

E_{Total} is the total energy of the system. The larger the K value, the stronger the ability of the target to absorb the impact energy.

5.4.2 Finite Element Model

The finite element model and meshing are shown in Fig. 5.15. This work only considered the single shot peening, which is the basis for the simulation of shot peening and strengthening process simulation. The shot radius is $R = 0.4$ mm, and the target size is $10R \times 10R \times 5R$ to weaken the influence of boundary effect. In order to simplify the calculation, according to the symmetry, 1/4 specimens were analyzed. The contact surface between the shot and the target is defined as the hard contact in normal direction, and the tangential direction is defined as the frictionless contact mode. The two symmetry planes and the bottom surface were constrained in their normal directions. The incident direction is perpendicular to the incident ($\alpha = 90°$), and the shot is only allowed to move freely in the direction 2 (refer to the coordinate system).

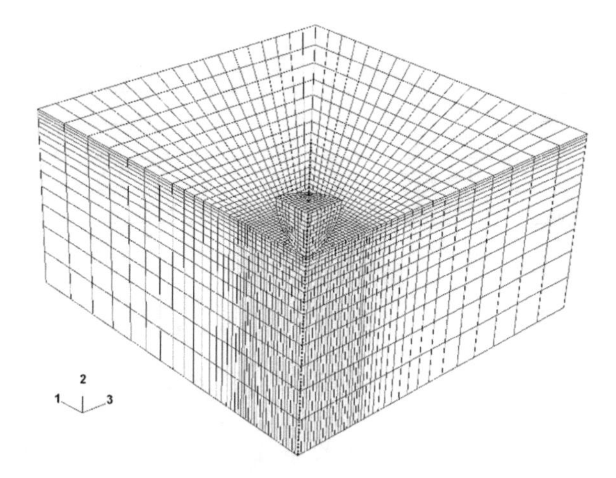

Fig. 5.15 Shot peening finite element model

This section is concerned with the problem of energy conversion at the moment of collision, so that the effect of damping is neglected. The shot and the target are in contact at the beginning of the analysis step. The analysis time is $t = 4 \times 10^{-5}$ s. It is proved that the kinetic energy of the system tends to be stable at the end of the calculation.

The shot is made of cast steel; its density is $\rho = 5500\ \text{kg/m}^3$, which is simplified as a rigid body. For the target, its density is $\rho = 7800\ \text{kg/m}^3$, the Young's modulus of E varied between 50–250 GPa, the Poisson's ratio $\upsilon = 0.3$, assuming that the target yield obeys the Mises yield criterion, and the yield strength of σ_y changed between 200 and 600 MPa. The plastic deformation stage is assumed to be linear strain hardening, and the strain hardening rate H take 100, 800, and 1500 MPa, respectively.

The shot exerts an impact load on the target at the moment of impact, which can be achieved by giving mass and initial velocity of the shot. The shot is rigid body, its mass is calculated according to its density and radius, and then it is defined in the form of point mass on the shot rigid body. The initial velocity of the shot is also defined on the rigid reference point, and $v_i = 100$ m/s.

5.4.3 *Model Verification*

The research on the shot peening test of 40Cr steel was carried out by Gao et al. [21]. It was found that the residual stress field of shot peening was related to the mechanical properties of the target and the shot peening process parameters; the following relations are established between the two:

Table 5.4 Comparison of the results of the simulation with Gao [21]

	Simulation results presented in this paper	Predicted value	Error (%)
σ_{mrs}	698 MPa	624 MPa	10.6
σ_{srs}	427 MPa	508 MPa	15.9
Z_0	0.47 mm	0.51 mm	8.5
Z_m	0.15 mm	0.14 mm	6.7

$$\sigma_{mrs} = 0.86\sigma_y - 51 \tag{5.19}$$

$$\sigma_{srs} = m(114 + 0.563\sigma_y)(m = 0.99\text{–}1.1) \tag{5.20}$$

$$Z_0 = (1.41D_d - 0.18R)\left[1 + 0.09(C - 1)^{0.55}\right] \tag{5.21}$$

$$Z_m \approx 0.28Z_0 \tag{5.22}$$

The mechanical parameters of 40Cr steel are input into the finite element model to calculate, and the residual stress field parameters are compared with the predicted values of the upper form (see Table 5.4). It can be seen that the calculated results of the model are in good agreement with the predicted values, which proved that the model is suitable and effective.

5.4.4 Effect of Young's Modulus of Target

In the analysis of this section, the effects of vibration and surface heat on energy dissipation are not considered, so the energy of the system can only be transformed between the kinetic energy of the shot and the strain energy of the target. The effect of the Young's modulus of the target on the kinetic energy, energy conversion ratio, and residual stress field of the system is shown in Fig. 5.16. Figure 5.16a shows that with the increase of the Young's modulus of the target, the kinetic energy of the system increases with the decrease of the kinetic energy at the input stage, while the kinetic energy is reduced in the elastic recovery stage. That is, with the increase of Young's modulus of the target, the energy absorption capacity is enhanced, and the K value of the energy conversion rate increases. It can be seen from Fig. 5.16b that the energy conversion rate of K and the Young's modulus of the target are nonlinear and monotonically increasing.

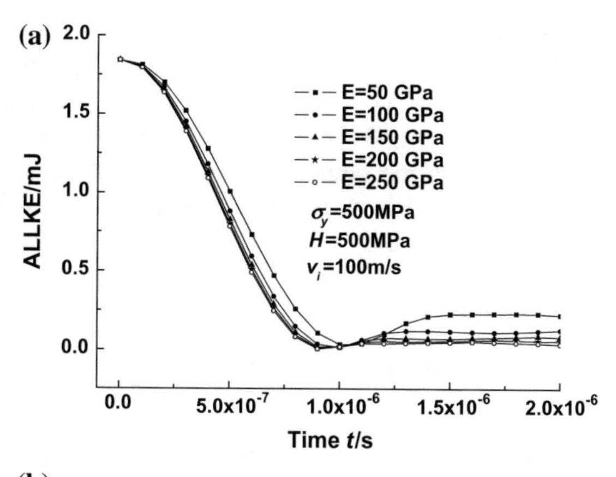

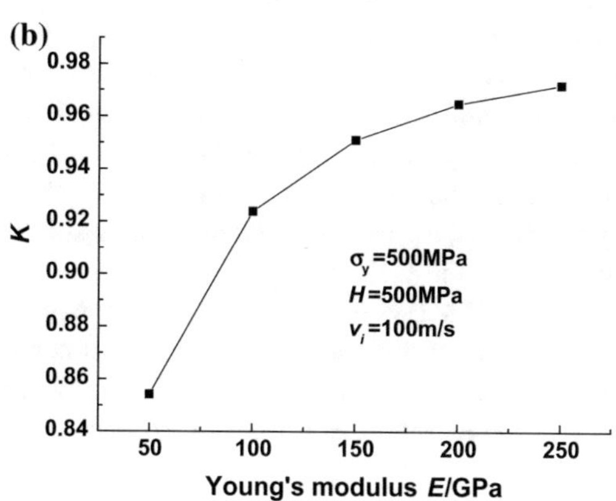

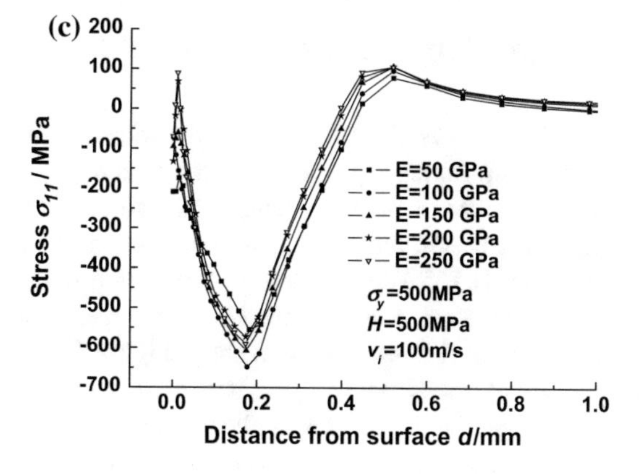

Fig. 5.16 Effect of Young's modulus on energy conversion and residual stress distribution

Figure 5.16c shows the variation of transverse residual stress in depth direction under different Young's modulus. As can be seen from Fig. 5.16c, with the increase of the Young's modulus of the target, the surface residual compressive stress decreases and converts to be residual tensile stress. When the Young's modulus of the target is in the range of 50–100 GPa, the maximum residual stress increases firstly with the increase of Young's modulus and then decreases with the increase of Young's modulus.

5.4.5 *Effect of Target Yield Strength*

Figure 5.17a shows the variation of the kinetic energy of the system with time under different yield strengths. It can be seen that with the increase of the yield strength of the target, the kinetic energy decay rate of the system is accelerated at the incident stage. In the elastic recovery stage, the kinetic energy increases with the increase of the yield strength of the target. As can be seen from Fig. 5.17b, the K value monotonically decreases linearly with the increase of the yield strength of the target, that is, the ability of the target to absorb the impact energy decreases with the increase of the yield strength.

Figure 5.17c shows the variation of the transverse residual stress along the depth of the target under different yield strengths. It can be seen that with the increase of the yield strength of the target, the maximum residual compressive stress increases, and the residual compressive stress field becomes shallower.

5.4.6 *Effect of Strain Hardening Rate*

Figure 5.18a shows the variation of the kinetic energy of the system with time under different hardening rates. It can be seen that the kinetic energy decay rate increases with the increase of hardening rate at the incident stage, and the kinetic energy of the elastic recovery phase increases in the elastic recovery stage. As can be seen from Fig. 5.18b, the energy conversion rate of K decreases monotonically linearly with the increase of strain hardening rate, that is, the ability of the target to absorb the impact energy is weakened. Figure 5.18c gives the variation of transverse residual stress with depth. It can be seen that with the increase of strain hardening rate, the surface residual compressive stress decreases and converts to be residual tensile stress. The maximum residual compressive stress and residual compressive stress field depth increase with the increase of strain hardening rate.

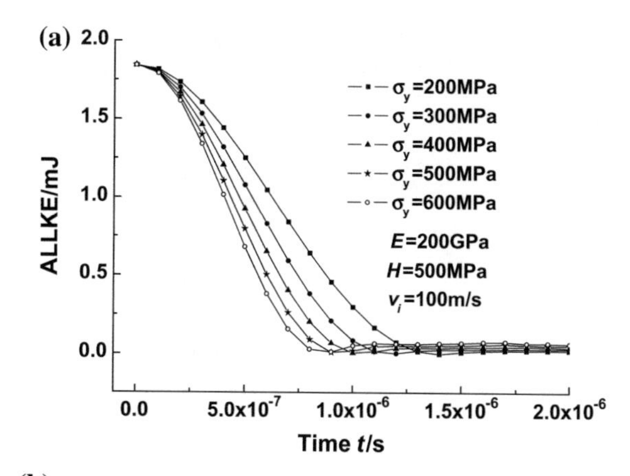

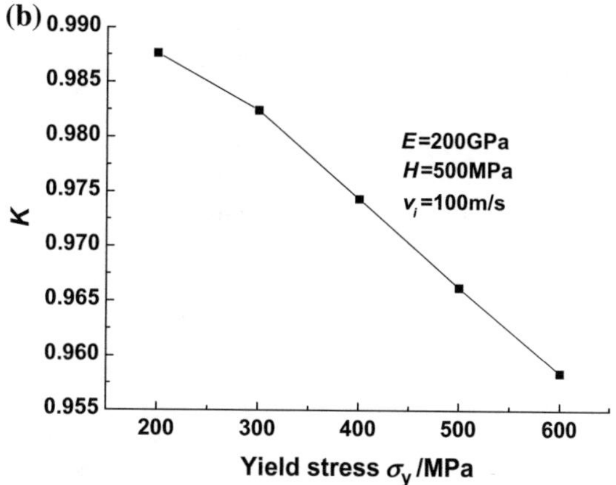

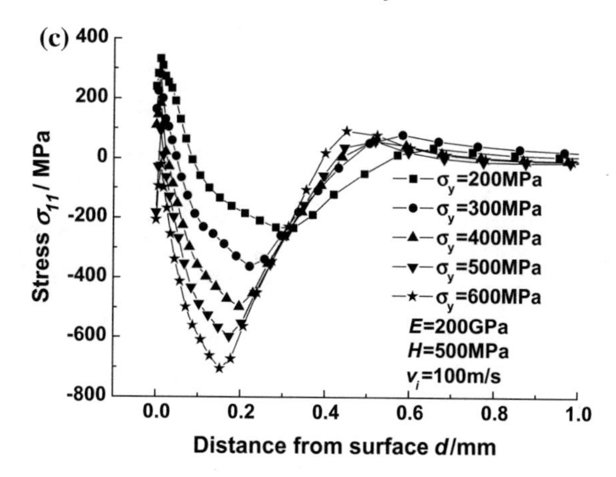

Fig. 5.17 Effect of target yield strength on energy conversion and residual stress distribution

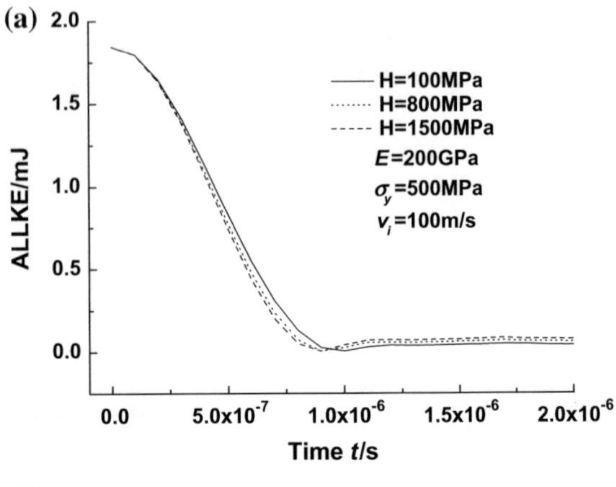

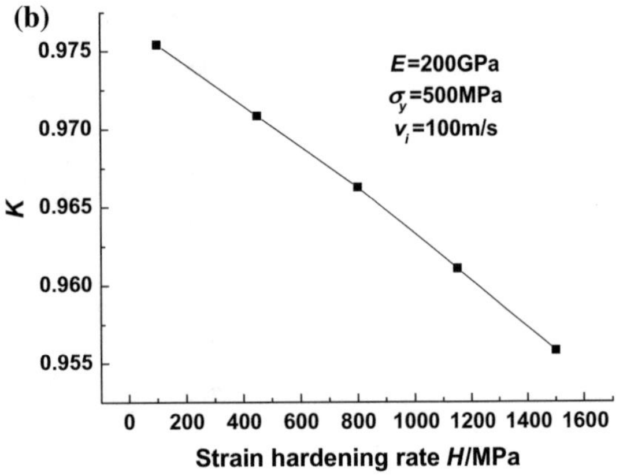

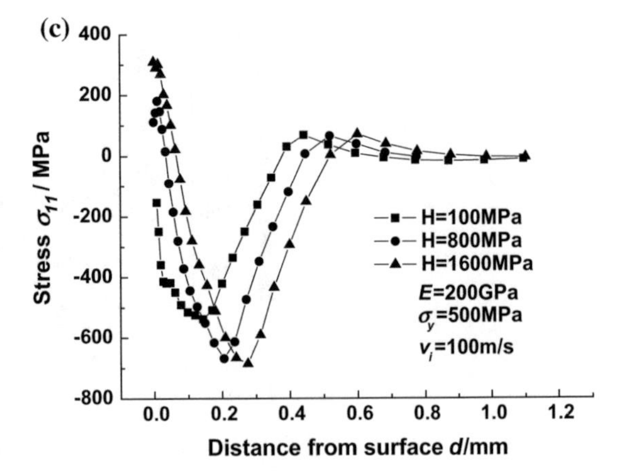

Fig. 5.18 Effect of strain hardening rate on energy conversion and residual stress distribution

References

1. Editorial board of aircraft design manual. 2000. *Aircraft design manual (structural design)*. Aviation Industry Press.
2. HB/Z 26-92. 1992. General specification for shot peening process of aviation parts.
3. Renzhi, Wang. 2011. *The research proceedings of metal material shot peening strengthening and surface integrity*. Beijing: China Astronautic Publishing House.
4. Fonte, M., F. Romeiro, and M. Freitas. 2007. Environment effects and surface roughness on fatigue crack growth at negative ratios. *International Journal of Fatigue* 29: 1971–1977.
5. Yukui, Gao. 2010. Effect of shot peening strengthening on microstructure of TC4 titanium alloy. *Rare Metal Materials and Engineering* 39 (9): 1536–1539.
6. Child, D.J., G.D. West, and R.C. Thomson. 2011. Assessment of surface hardening effects from shot peening on a Ni-based alloy using electron backscatter diffraction techniques. *Acta Materialia* 59: 4825–4834.
7. Dalaei, K., B. Karlsson, and L.-E. Svensson. 2010. Stability of residual stresses created by shot peening of pearlitic steel and their influence on fatigue behaviour. *Procedia Engineering* 2: 613–622.
8. Dalaei, K., B. Karlsson, and L.-E. Svensson. 2011. Stability of shot peening induced residual stresses and their influence on fatigue life time. *Materials Science and Engineering A* 528: 1008–1015.
9. Gangaraj, S.M.H., Y. Alvandi-Tabrizi, G.H. Farrahi et al. 2011. Finite element analysis of shot-peening effect on fretting fatigue parameters. *Tribology International* 44: 1583–1588.
10. Yukui, Gao. 2007. The relaxation law of residual stress on the surface of ultra high strength steel shot peening. *Journal of Material Heat Treatment* 28: 102–105.
11. Liu, W.C., J. Dong, P. Zhang, et al. 2011. Improvement of fatigue properties by shot peening for Mg-10Gd-3Y alloys under different conditions. *Materials Science and Engineering A* 528: 5935–5944.
12. Xiang, Ling, Peng Weiwei, and Ni Hongfang. 2006. Finite element simulation of 3D residual stress field of shot peening. *Chinese Journal of Mechanical Engineering* 42 (8): 182–189.
13. Bo, Chen, Gao Yukui, Wu Xueren, Ma Shaojun. 2010. Small crack behavior and life prediction of shot peening 7475-T7351 aluminum alloy. *Chinese Journal of Aeronautics* 33 (3):519–525.
14. Yonghong, Yang, Qiao Mingjie, and Zhang Weihong. 2009. The influence law of shot peening condition on residual stress field. *Surface Engineering of China* 22 (2): 45–48.
15. Hongwei, Zhang, Zhang Lidu, and Wu Qiong. 2010. Three dimensional numerical analysis of residual stress field of shot peening. *Journal of Aerospace Power* 25(3):603–609.
16. Yukui, Gao. 2003. Characteristics of residual compressive stress field of high strength steel shot peening. *Metal Heat Treatment* 28 (4): 42–44.
17. Gumao, Mi. 1983. *The formation and Countermeasures of residual stress*. Beijing: China Machine Press.
18. Weichang, Qian. 1984. *Mechanics of armour piercing*. Beijing: National Defence Industry Press.
19. Carvalhoa, A.L.M., and H.J.C. Voorwaldb. 2009. The surface treatment influence on the fatigue crack propagation of Al 7050-T7451 alloy. *Materials Science and Engineering A* 505: 31–40.
20. Pilipenko, A. 2001. *Computer simulation of residual stress and distortion of thick plates in multi-electrode submerged arc welding*. PH. D thesis. Department of March. Designing and Material Technology, Norwegian University of Science and Technology.
21. Gao, Y.K., M. Yao, and J.K. Li. 2002. An analysis of residual stress fields caused by shot peening. *Metallurgica and Materials Transactions A* 33A (6): 1775–1778.

Chapter 6
Anti-fatigue Design and Analysis of Joints

Aircraft wings consist a variety of connection joints, which are often the most dangerous places where strength failure or fatigue damages occur. Although people are trying to develop the integrated structure of the aircraft to avoid the joints, a large number of aircraft structures are still using the traditional form of lap joints to realize the transmission and distribution of load among different parts due to the requirements of design, process, and economic constraints [1]. Different parts, such as body skins, wing panels, etc., often use different forms of joints. The fatigue quality of the joints depends on the type of fasteners, the form of assembly, the quality of the hole, the size of the hole diameter, and other factors [2].

6.1 Single Shear Lap Joints

The scarfed lap joint structure is a way of connecting the airplane panels, in which the overlap angle is introduced on the faying surface of the lap. The scarfed panels are connected by 3–5 rows of rivets or bolts as shown in Fig. 6.1. Compared with the flat plate connection, this docking scheme has the advantages of lightweight, averaged pin load distribution, and low local stress around the fastening holes, thereby improving the life of the connecting parts.

6.1.1 Fatigue Test

6.1.1.1 Size of Single Shear-Tapered Test Piece

The material used for the specimen is an aluminum alloy 2024, the lap is inclined, the two pieces are assembled with the countersunk head screw into the lap joints specimen. After the lap, the total thickness is 8 mm. The typical geometrical

J. Liu et al., *Long-Life Design and Test Technology of Typical Aircraft Structures*,
https://doi.org/10.1007/978-981-10-8399-0_6

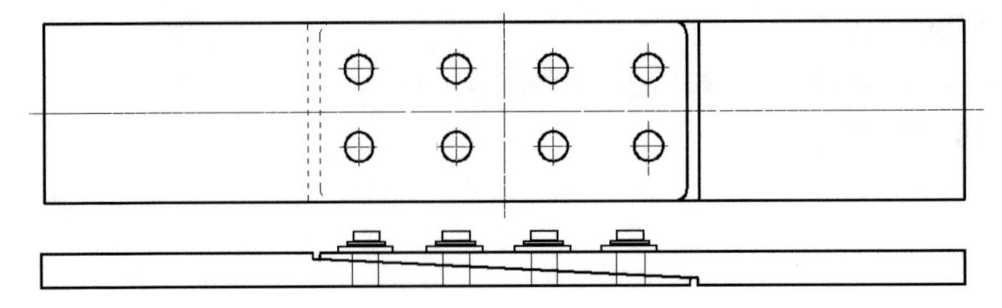

Fig. 6.1 Scarfed lap joints

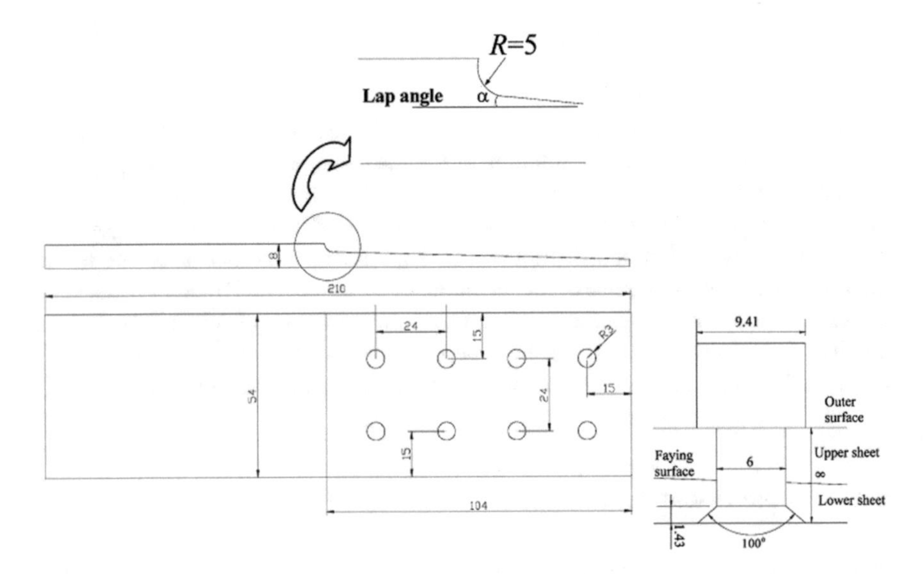

Fig. 6.2 Geometries of the lap sheet and the countersunk rivet (unit: mm)

dimensions are shown in Fig. 6.2. The screw preload is 5.5 kN and the interference is 0.3%. Countersunk screws are made of titanium alloy. Typical mechanical properties of the material are calibrated by tensile tests, as shown in Table 6.1.

Table 6.1 The chemical composition of aluminum alloy 2024 used in this work

Composition	Al	Cu	Mg	Mn	Si	Fe	Cr	Zn
Mass fraction	92.13%	4.41%	1.52%	0.59%	0.5%	0.5%	0.1%	0.25%

6.1.1.2 Fatigue Test and Its Results

According to the overlap angle and the number of screw rows, the test pieces are divided into 6 groups, each group has 7 pieces. All specimens were carried out on a servo hydraulic fatigue tester INSTRON8802. Anti-bend clamp was applied on both outer surfaces of the test piece to prevent out-of-plane bending (see Fig. 6.3). The maximum applied fatigue load is F_{max} = 30 kN, the stress ratio is R = 0.06, and the loading frequency is 4 Hz.

The fatigue life of each specimen is shown in Table 6.2. It can be seen from Table 6.2 that the fatigue life increases with the increase of the lap angle when the number of rows is the same, and the fatigue life of the five rows of nails is higher

Fig. 6.3 Fatigue test device

Table 6.2 Inclined structure test fatigue life

Specimen number	Rows of rivets	Lap angle (°)	Fatigue life (cycle)	Coefficient of variation
2110-003	4	1.68	411,196	0.23
2110-005	4	1.10	293,576	0.21
2110-007	4	0.55	215,279	0.26
2110-009	4	0	128,527	0.18
2310-003	5	1.39	613,804	0.30
2310-005	5	0.72	310,475	0.27

Fig. 6.4 Fatigue failure of the joint

than that of the four rows. The specimen is broken at the first row of holes in the lower deck and the typical fracture failure specimen is shown in Fig. 6.4.

6.1.1.3 Fracture Analysis

For the lap structure, the hole stress concentration is usually caused by the contact of the fastener and the hole and the secondary bending moment. There have also been studies showing that the coefficient of friction between the contact surfaces and the fastener preload force can lead to stress concentration in the hole. The crack of the fastening hole is usually nucleated at two positions where a crack is nucleated at the junction of the fastener and the cell wall, which is usually caused by local high stress; the other crack is inside the bore wall at the initiation, this crack is usually caused by fretting wear fatigue [3].

After the fatigue test, the fracture nucleation mechanism was studied by stereoscopic optical microscopy. As can be seen from Fig. 6.5, due to changes in the overlap angle, the crack is initiated at two locations.

When the overlap angle $\propto = 0^\circ$ (flat lap), the crack is nucleated at the edge of the fastening hole (see Fig. 6.5a). The crack nucleation is caused by stress concentration.

With the increase of the overlap angle, the crack nucleation takes place at the hole wall and has a short distance from the overlapping surface (see Fig. 6.5b), accompanied by slight friction damage. This phenomenon is more pronounced when the overlap angle $\propto = 1.68^\circ$ (Fig. 6.5c), and the crack initiation position is at a certain distance from the edge of the hole in the overlap surface. It can be seen from the fracture analysis that the overlap angle has some influence on the initiation position and initiation mechanism of the crack initiation.

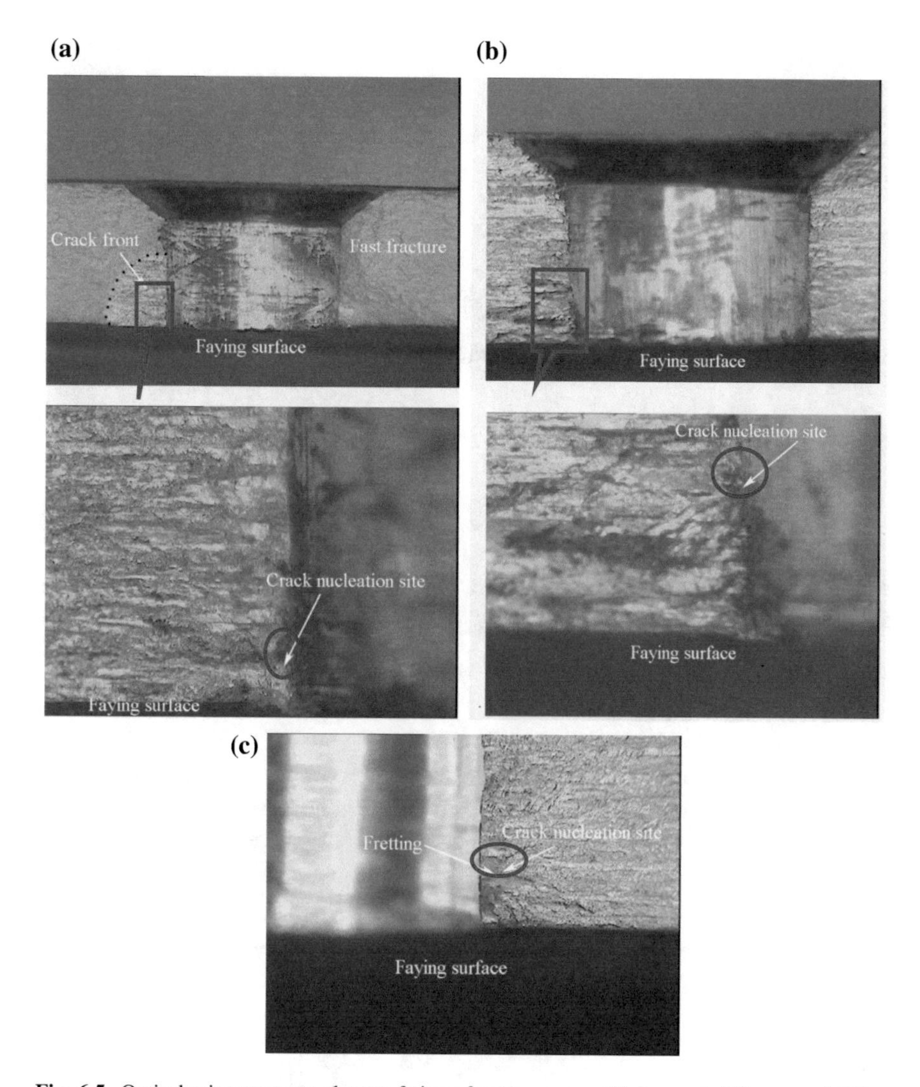

Fig. 6.5 Optical microscopy to observe fatigue fracture, **a** $\propto = 0°$, **b** $\propto = 1.12°$, **c** $\propto = 1.68°$

6.1.1.4 Model Establishment

The scarfed lap finite element model and mesh are shown in Fig. 6.6. In the finite element model, the mesh of the hole is refined. The model size is the original size of the specimen (see Fig. 6.1).

For all contact surfaces, the friction coefficient is set to $\mu = 0.3$ and the slip equation is set to small slip. In the plastic range, the plastic flow behavior follows the nonlinear stress–strain curve as follows:

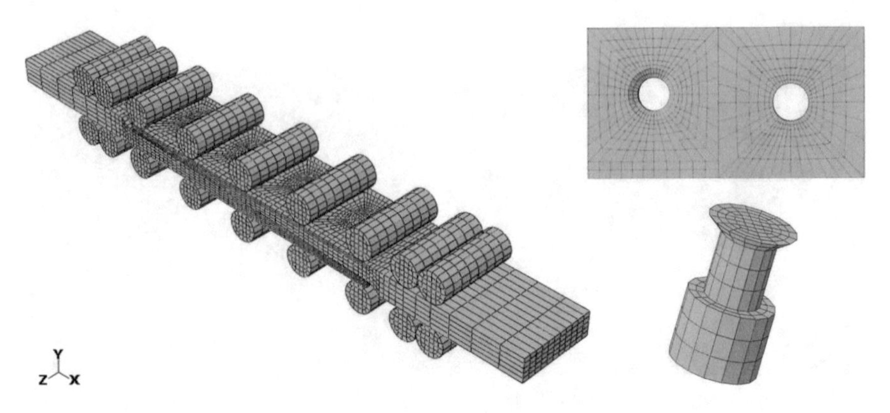

Fig. 6.6 Fractal finite element model

$$\sigma = C\varepsilon^n \tag{6.1}$$

where σ is the flow stress, ε is the strain, C is the strength coefficient, and n is the strain hardening index.

6.1.1.5 Model Validation

The static test was carried out to verify the model. Three static tests were carried out on the specimen. Strain gauges-pasted positions on strain test specimens are shown in Fig. 6.7. The measured strain is compared with the finite element calculation

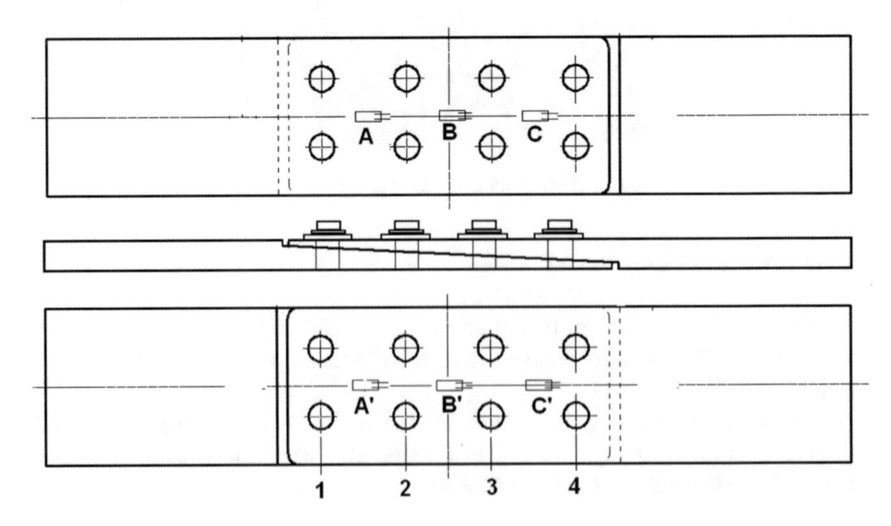

Fig. 6.7 Strain gauge distribution

Table 6.3 Comparison of test strain and finite element calculation results (unit: 1.0×10^{-6})

Specimen number	Measuring point A	Measuring point B	Measuring point C	Measuring point A′	Measuring point B′	Measuring point C′
1	1793	1338	760	768	1280	1867
2	1172	1125	1027	1113	1183	1314
3	1270	1088	1016	1022	1204	1315
Mean	1412	1184	934	967	1222	1498
Finite element results	1483	1104	878	873	1134	1600
Error	5.05%	6.75%	6.03%	9.85%	7.21%	6.8%

results as shown in Table 6.3. The maximum relative error between the test strain and the finite element calculation is 9.85%, which shows the effectiveness of the scarfed lap finite element model.

6.1.1.6 Impact of Rivet Rows on Pin Load

The number of rivets is another important design detail for lap joint. Figure 6.8 shows the effect of the number of rows of rivets on the pin load distribution and the maximum Mises stress around the holes. From Fig. 6.8, it can be seen that when the number of rows of rivets changes, the pin load of the first row is basically unchanged, but the proportion of the pin load in the middle row is changed. When the number of rows is four, the pin load distribution is more uniform, and the minimum pinning ratio is 9%, the maximum stress is larger (210 MPa), but still

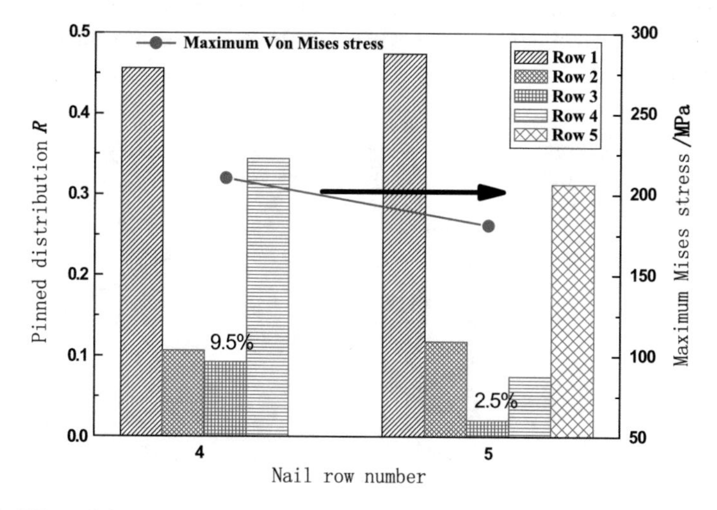

Fig. 6.8 Effect of rivet row number on pin load distribution and maximum Mises stress

within the allowable range. When the number of rows is five, the third row of rivets bears 2.5% of the overall load, resulting in waste of materials. Considering the design requirements, economy, weight saving and load transfer, the selection of four rows of rivets is more reasonable for the scarfed lap joints.

6.1.1.7 The Effect of the Overlap Angle on Pin Load

Stress concentration of shear joints leads to fatigue damage and crack initiation. The stress concentration position is related to the load transfer modes. This section considers the following two load transfer modes:

(1) rivet bearing shear force,
(2) friction between the upper and lower lap plate.

The first load transfer mode is related to factors such as scarfed angle, interference, and cold expansion. The second load transfer mode is mainly related to the rivet preload and the friction between the upper and lower lap plates. The pin load distribution can be expressed as follows:

$$R = p_t/(p - p_f), \quad (6.2)$$

where p_t is the load transferred by each row of rivets, p is the total load applied, and p_f is the interfacial friction force. The magnitude of the load bearing and the interplate friction can be obtained by the contact force output from the finite element calculation results. Taking the 4-row rivets as an example, Fig. 6.9 shows the variation of the pin load distribution with the scarfed angle. It can be seen that the first and the fourth row of rivets bear the maximum load, and they all decrease with the increase of the scarfed angle. On the other hand, the bearing loads of the third and fourth row of rivets are much smaller, and they all increase with the increase of the scarfed angle. The increase of the scarfed angle makes the distribution of the pin load more uniform.

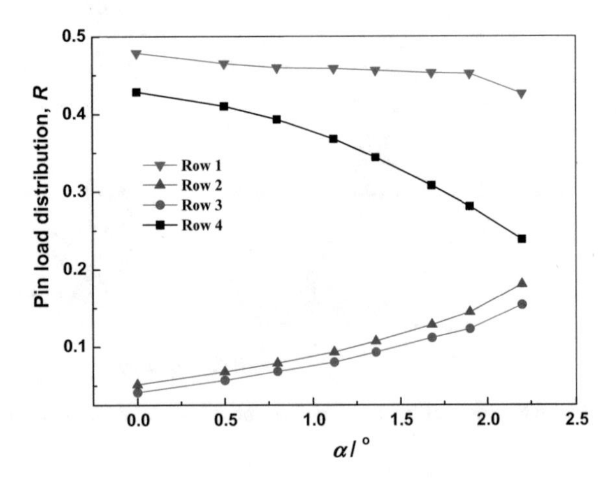

Fig. 6.9 Effect of lap angle on pin load distribution (four rows of nails)

6.1.2 *Analysis of Hole Stress*

6.1.2.1 Maximum Principal Stress Analysis

In general, the direction of crack growth is perpendicular to the direction of maximum principal stress. The maximum principal stress distributions of the upper and lower plate are shown in Fig. 6.10. It can be seen that the maximum principal stress occurs at the first row of holes in the lower plate, and the test results show that the specimen fractured at the same location (see Fig. 6.4).

The effect of the scarfed angle on the maximum principal stress is shown in Fig. 6.11. It can be seen that the stress level is the lowest when the lap angle $\alpha = 1.68°$.

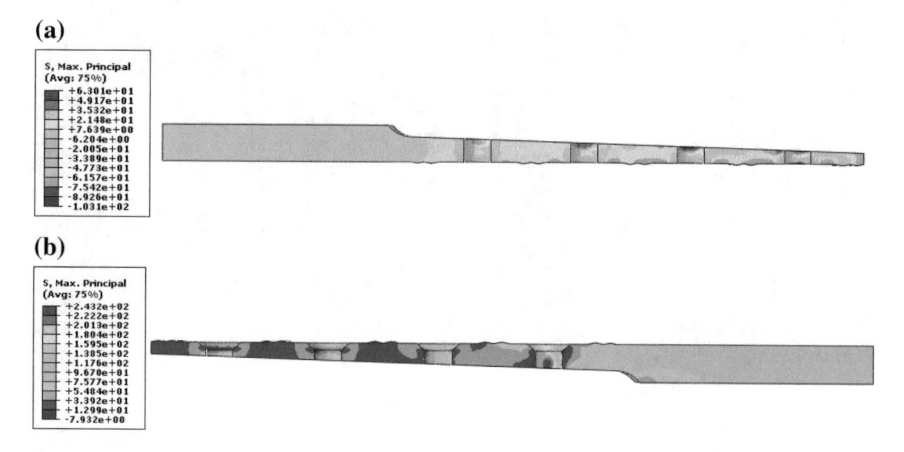

Fig. 6.10 Stress distribution on the symmetry plane of the mating sheet, **a** upper plate, **b** lower plate

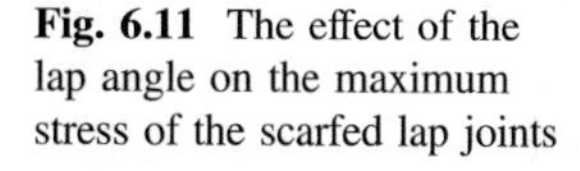
Fig. 6.11 The effect of the lap angle on the maximum stress of the scarfed lap joints

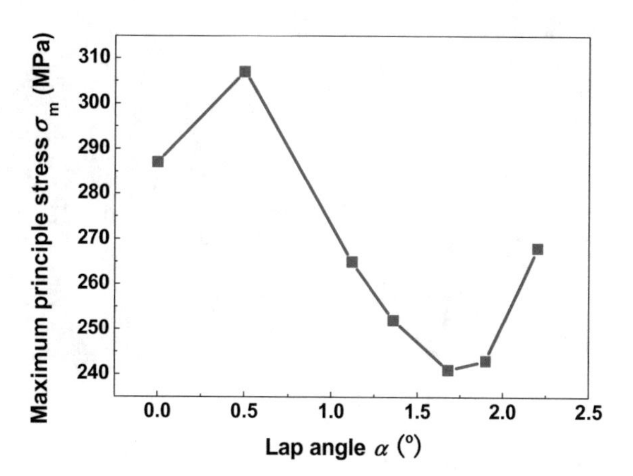

6.1.2.2 Analysis of the Circumferential Stress of the Fastening Hole

For fastening holes, crack always initiates and propagates in the radial direction (perpendicular to the circumferential stress). Therefore, the distribution of stress along the circumferential path is an important aspect of our concern. Figure 6.12 shows the circumferential distribution of the first row of holes in the lower plate. It can be seen that the effect of the scarfed angle on the circumferential stress of the hole is remarkable. Figure 6.13 shows that the maximum tensile stress occurs in the direction of $\theta = 90°$ and 270°. The hole stress is the smallest when the lap angle $\alpha = 1.68°$.

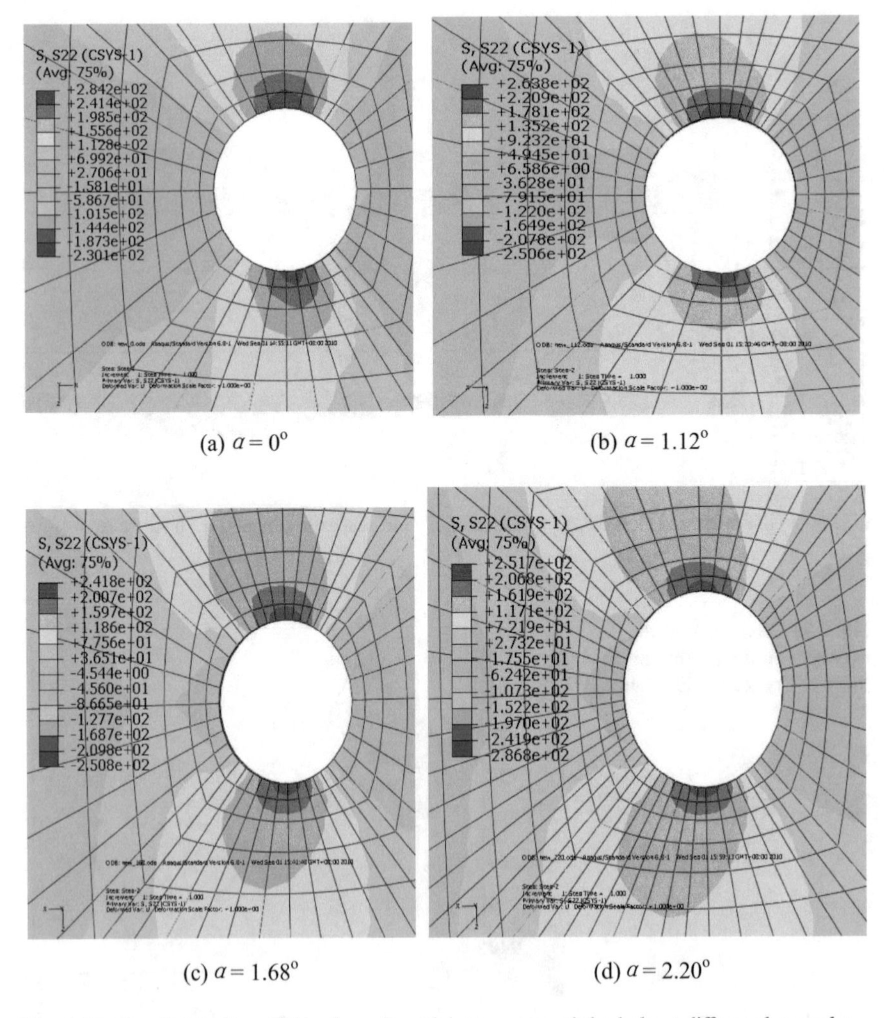

(a) $a = 0°$ (b) $a = 1.12°$

(c) $a = 1.68°$ (d) $a = 2.20°$

Fig. 6.12 The distribution of the circumferential stress around the hole at different lap angles

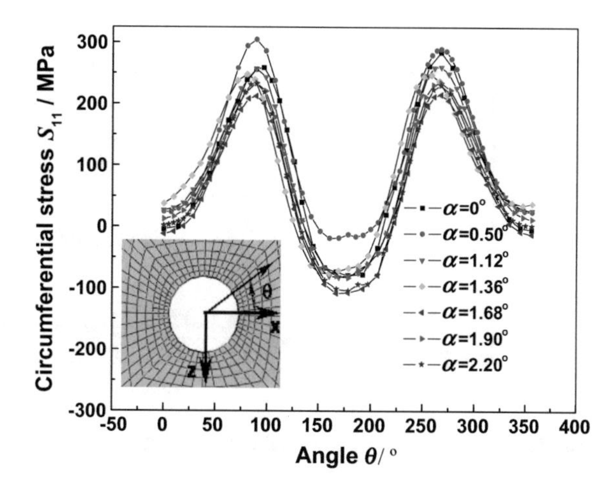

Fig. 6.13 Effect of lap angle on tensile stress around holes

6.2 Double Shear and Interference Fit Joints

In the aircraft structure, the bolts that work under shear force account for a large proportion, in which the parts and bolts are sheared to transfer the outer load, such as the wing and the fuselage connections. In addition, in the aerospace structure, fastening holes often use an interference fit to improve the fatigue life and reliability of the connector. This is because the interference fit can produce self-phase equilibrium residual compressive stress [4] on fastening hole wall, reducing the hole stress amplitude of the fatigue load spectrum.

6.2.1 Effect of Interference on Pin Load

6.2.1.1 Double Shear Connector Geometry

The double shear lap joint analyzed in this section is divided into two types: interference fit (1.0% interference) and precision bolt according to the type of fasteners. Figure 6.14, the total length of the specimen is 290 mm, the thickness of the upper and lower panels are 5 mm, the thickness of the middle plate is 6 mm, the margin to the center of fastening hole is 20 mm, the end distance to the fastening hole is 15 mm. The diameter of the fastening hole is 6 mm, the hole spacing is 30 mm. The three plates are connected with 90° countersunk bolts. In order to avoid out-of-plane bending during the loading process, two pads were added between the two pieces of test pieces to ensure that the loading center is in the central line of the specimen.

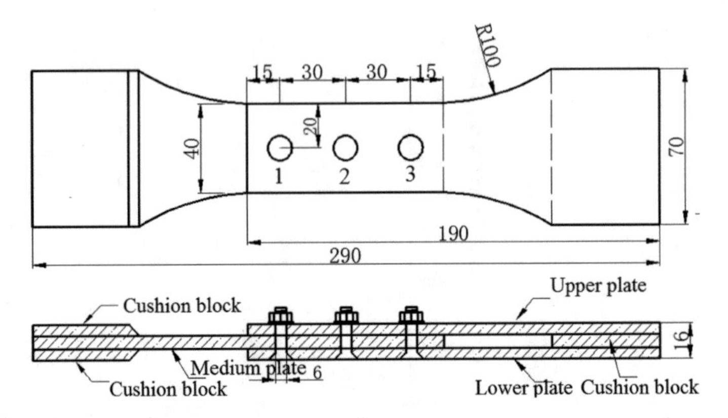

Fig. 6.14 Geometric model of double shear specimen

Table 6.4 Mechanical properties of materials of non-transfer fastening holes

	Material grade	*E*/GPa	μ
Lap	7050	71	0.33
Bolt	30CrMnSiA	196	0.30

The upper plate, the middle plate, and the lower plate are made of aluminum alloy 7050. The bolt material is the alloy structural steel 30CrMnSiA, and the typical mechanical properties of each material are obtained by the tensile test as shown in Table 6.4.

6.2.1.2 Bolt Load Test

Figure 6.15 shows the locations of strain gauges on the specimen. The locations of 1, 2, 3 are the bolt locations. The locations of A, B, C are the strain gauge locations. The multi-channel dynamic strain measurement acquisition instrument DH3817 was used to record data.

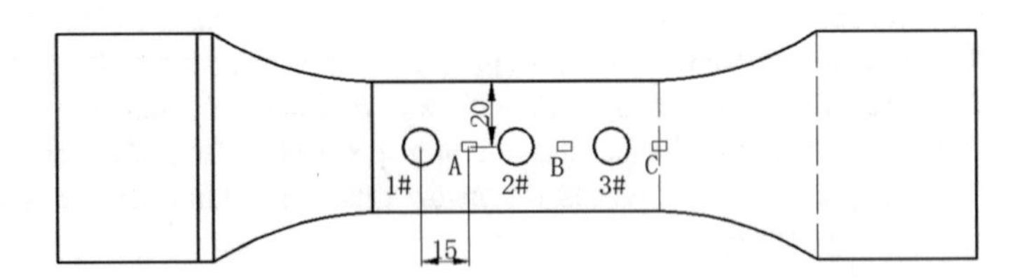

Fig. 6.15 Strain guage position diagram

The bolt load can be obtained by the static test. The static test of the double shear connection is carried out on the hydraulic servo fatigue testing machine Instron 8802 as shown in Fig. 6.16. Displacement control loading method is used with a loading rate of 0.25 mm/min, and loading until the damage of test piece.

After the static damage test, the displacement load curve is obtained. As can be seen from Fig. 6.17, when the load is small, there is almost no difference between the two. With the increase of the load, the plastic deformation step is carried out.

Fig. 6.16 Static test of double shear joint

Fig. 6.17 Displacement load curves for different fasteners

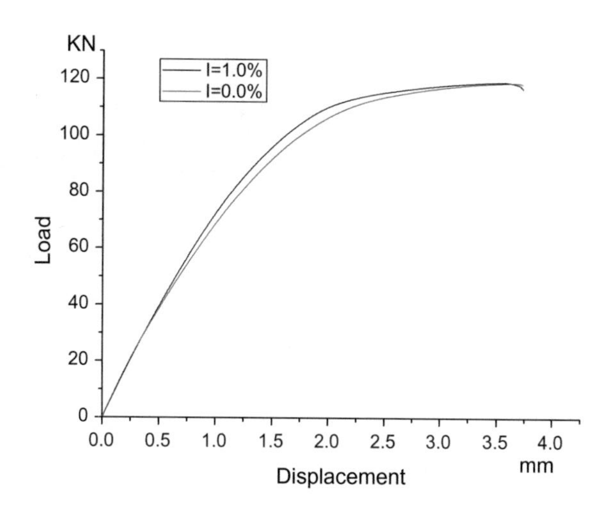

Table 6.5 Pin load assignment

Nail row	Precision fit bolts connection (%)	Interference fit bolts (%)
1	43.0	41.4
2	22.7	23.1
3	34.3	35.5

The overall stiffness has a more obvious increase, into the destruction phase the static strength of the two are almost the same.

Table 6.5 shows that the ratio of the load to the first row of nails is the largest, and the maximum nail for the interference fit bolts is the maximum. The ratio is slightly lower than the precision with the bolt.

6.2.1.3 Finite Element Model of Double Shear Joint and Its Verification

The three-dimensional finite element model was established by ABAQUS to simulate the transmission load and the stress distribution of the double—the finite element model is shown in Fig. 6.18.

The preload of the bolt is taken into account in the analysis. In order to facilitate convergence, first we will use a smaller amount of interference, and then increased to the amount of interference to solve. In order to accurately simulate the tensile stress process of the double shear joint, the whole finite element analysis process is divided into four steps: (1) bolt preload loading; (2) smaller interference amount; (3) actual interference amount; (4) apply a farfield uniform load. According to

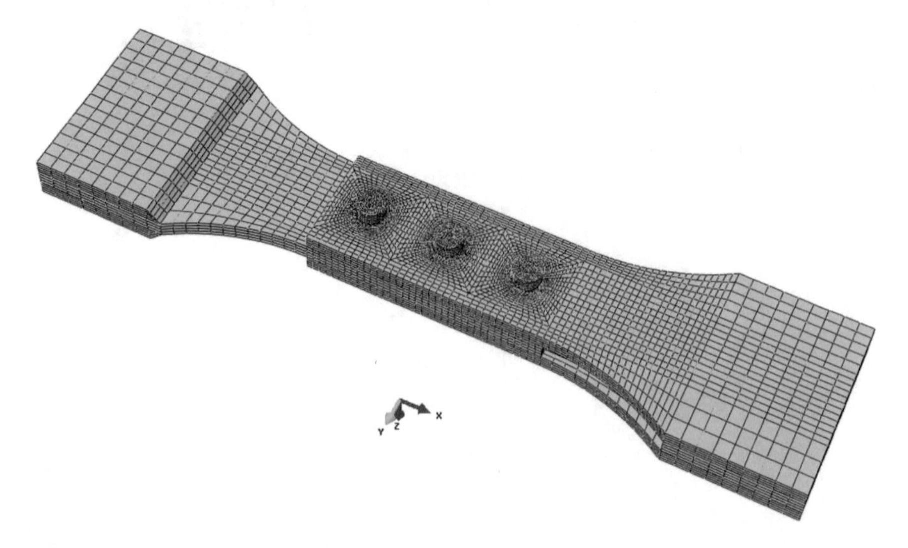

Fig. 6.18 Finite element model and mesh division

Table 6.6 Comparison of finite element analysis and test values for the pinned distribution

Nail row	Precision fit bolts connection		Interference fit bolts	
	Test results (%)	Finite element results (%)	Test results (%)	Finite element results (%)
1	43.0	41.5	41.4	47.6
2	22.7	27.7	23.1	25.6
3	34.3	30.8	35.5	26.8

Eq. (6.2), we can get the finite element results of the pinned distribution under each interference amount as shown in Table 6.6.

The maximum error of the calculated and finite element values is 8.7%, indicating that the established finite element model is more accurate and can be used for further stress analysis.

6.2.2 *Fatigue Test and Analysis*

6.2.2.1 Fatigue Test and Results

The double shear connectors for fatigue testing are divided into two types: the precision bolts are connected and the interference is 1% of the interference connection, for each of the 7 pieces. Fatigue tests were carried out at room temperature. Stress control was used under a tension-to-tension mode with a sine wave at a stress ratio R = 0.1 and a frequency of 5 Hz. The maximum load was 50KN.

The fatigue life of each specimen is shown in Table 6.7.

Using the pairwise contrast test method for data processing [5]. In this method, the number of pairs of children shall not be less than five pairs, in order to ensure that the difference has a strong ability to distinguish, the work of the test piece logarithm of 7 pairs should meet the applicable conditions.

In the data processing, the fatigue life of the test is taken first, N_{1i} is the fatigue life of the precision fitting bolts, N_{2i} is the fatigue life of the interference fit bolts, so

Table 6.7 Fatigue life of test specimen

Connection method	1	2	3	4	5	6	7	Average life
Precision fit	207,320	106,668	125,972	276,343	209,058	546,572	107,676	166,990
Interference fit	655,292	277,416	689,351	816,322	827,627	316,578	628,592	601,597

Table 6.8 Pairs of shear test pieces of fatigue life pairs of comparative analysis

Number	Project					
	Precision joint bolts specimen life N_{1i}	Interference fit bolts specimen life N_{2i}	Precision fitting bolts test piece life span X_{1i}	Interference fit bolts specimen logarithmic life X_{2i}	Logarithmic life difference Y_i	Y_i^2
1	207,320	655,292	5.3166	5.8164	0.4998	0.2498
2	106,668	277,416	5.0280	5.4431	0.4151	0.1723
3	125,972	689,351	5.1003	5.8384	0.7382	0.5449
4	276,343	816,322	5.4414	5.9119	0.4704	0.2213
5	209,058	827,627	5.3203	5.9178	0.5976	0.3571
6	135,896	316,578	5.1332	5.5005	0.3673	0.1349
7	107,676	628,592	5.0321	5.7984	0.7662	0.5871
$\sum$					3.8546	2.2674

that $X_{1i} = \lg N_{1i}$; $X_{2i} = \lg N_{2i}$. The difference between the results is $Y_i = X_{2i} - X_{1i}$. The calculation is shown in Table 6.8.

(a) Constructing statistics

$$t = \frac{\overline{Y} - \mu_Y}{\frac{S}{\sqrt{n}}} \tag{6.3}$$

where $Y = \frac{1}{n}\sum_{i=1}^{n} Y_i$; $S = \sqrt{\frac{1}{n-1}\left[\sum_{i=1}^{n} Y_i^2 - n\overline{Y}^2\right]}$.

According to the definition of t distribution, we can see $t \sim t\ (n-1)$.

(b) Select the letter level $1 - \alpha$

Take $\alpha = 0.05$, $1 - \alpha = 0.95$

(c) Establish probabilistic conditions

$$P\{|t| \leq t_\alpha\} = 1 - \alpha \tag{6.4}$$

According to the degree of freedom $n - 1 = 7 - 1 = 6$ and $1 - \alpha = 0.95$, check the distribution table $t_\alpha = 2.4469$.

(d) Calculate the confidence interval

From the above probabilistic conditions, it is very important that the event $\{|t| \leq t_\alpha\}$ is the occurrence of the event $\left\{\left|\frac{\overline{Y}-\mu_Y}{\frac{S}{\sqrt{n}}}\right| \leq t_\alpha\right\}$. The inequality is transformed and the confidence interval of 100 × (1 − α)% of $\mu_{\overline{Y}}$ is

$$\overline{Y} - t_\alpha \frac{S}{\sqrt{n}} \leq \mu_Y \leq \overline{Y} + t_\alpha \frac{S}{\sqrt{n}} \tag{6.5}$$

Substituting Table 6.5 results

$$\overline{Y} = \frac{1}{n}\sum_{i=1}^{n} Y_i = \frac{3.8546}{7} = 0.5507 \tag{6.6}$$

$$S = \sqrt{\frac{1}{n-1}\left[\sum_{i=1}^{n} Y_i^2 - n\overline{Y}^2\right]} = \sqrt{\frac{1}{7-1}[2.2674 - 7 \times 0.5507^2]} = 0.1552 \tag{6.7}$$

$$\overline{Y} - t_\alpha \frac{S}{\sqrt{n}} = 0.5507 - 2.4469 \times \frac{0.1552}{\sqrt{7}} = 0.4072 \tag{6.8}$$

$$\overline{Y} + t_\alpha \frac{S}{\sqrt{n}} = 0.5507 + 2.4469 \times \frac{0.1552}{\sqrt{7}} = 0.6942 \tag{6.9}$$

The 95% confidence interval of $\mu_{\overline{Y}}$ is

$$0.4072 \leq \mu_Y \leq 0.6942$$

Since the 95% confidence interval of $\mu_Y = \mu_{X_2} - \mu_{X_1}, \mu_{X_2} - \mu_{X_1}$ is

$$0.4072 \leq \mu_{X_2} - \mu_{X_1} \leq 0.6942$$

Because $\mu_{X_2} - \mu_{X_1} = \lg \mu_{N_2} - \lg \mu_{N_1} = \lg \frac{\mu_{N_2}}{\mu_{N_1}}$

Substituting the previous confidence interval expression, take

$$0.4072 \leq \lg \frac{\mu_{N_2}}{\mu_{N_1}} \leq 0.6942$$

Take the opposite number

$$2.5539 \leq \frac{\mu_{N_2}}{\mu_{N_1}} \leq 4.9454$$

Therefore, 95% of the grasp concluded that the life of the interference with the connector is the average life of the bolts with the average life of 2.5539–4.9454 times.

6.2.2.2 Fracture Analysis

After the test is completed, the fracture of all the test pieces is observed and analyzed. The fracture position of the test piece occurs at 1 # nail hole of the middle plate. As shown in Fig. 6.19, the typical fracture is shown in Fig. 6.20. Crack initiated at 1 # hole wall of middle plate, in the form of surface crack.

6.2.2.3 Stress Analysis

The stress distribution of the overlapped piece is shown in Fig. 6.21 by using the finite element model of the double shear joint established in the previous section. It can be seen that the maximum stress occurs at the 1 # fastening hole in the middle plate, which is consistent with the experimental fracture site and again the accuracy of the model.

The effect of the amount of interference on the maximum stress is shown in Fig. 6.22. It can be seen that the interference fit can effectively reduce the main stress of the hole edge. When the interference amount is 1%, the maximum stress of

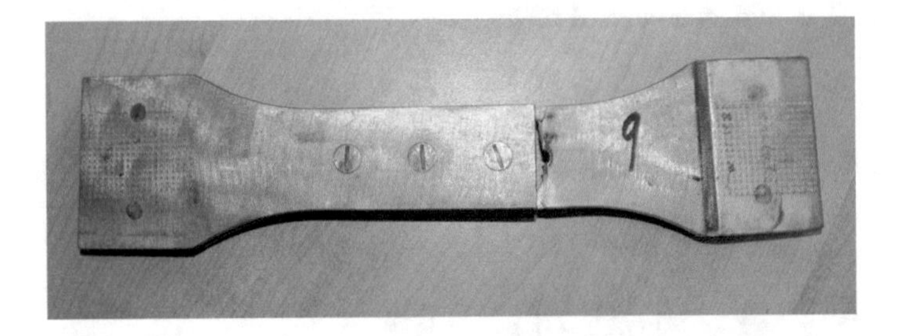

Fig. 6.19 Fracture position

Fig. 6.20 Typical fracture morphology

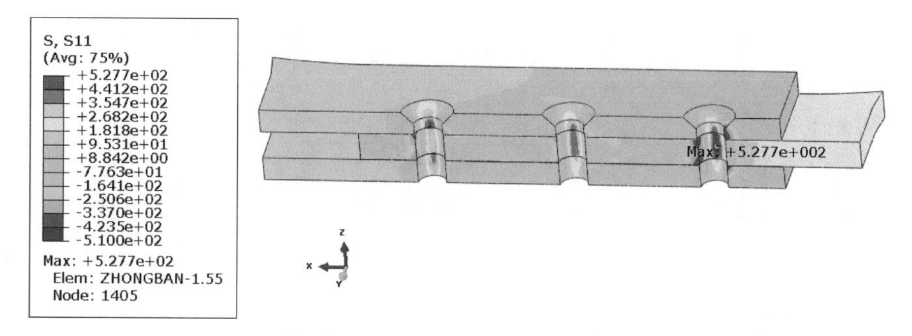

Fig. 6.21 Pinhole stress distribution

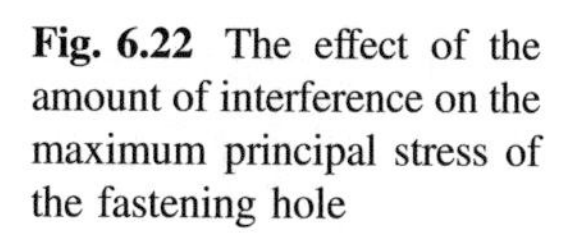

Fig. 6.22 The effect of the amount of interference on the maximum principal stress of the fastening hole

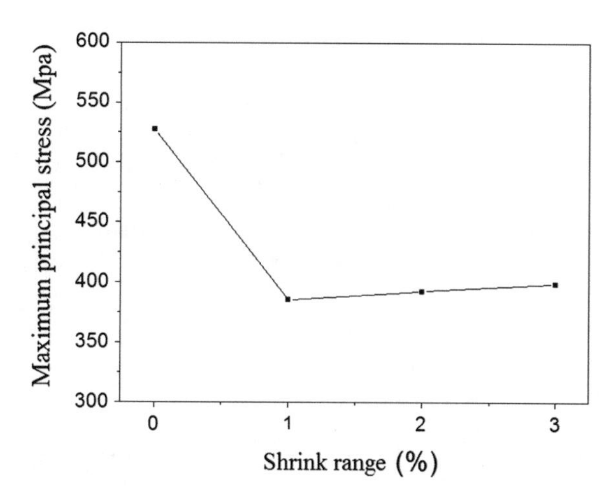

the fastening hole is the smallest. In other words, increasing interference amount does not always get better results. There is a best interference value, larger or smaller this value, the maximum stress of hole increases.

6.3 Reverse Double Dogbone Joints

A thin structure on the aircraft, usually using the head screw connection, that is, the head and the outer surface flush connection, which effectively avoid the ordinary connection due to the screw exposed to the outer surface of the aircraft on the aerodynamic characteristics of the impact. In the engineering test, when the simulation of low-load transfer screw connection, the general use of double canine bone specimens [6]. As shown in Fig. 6.23.

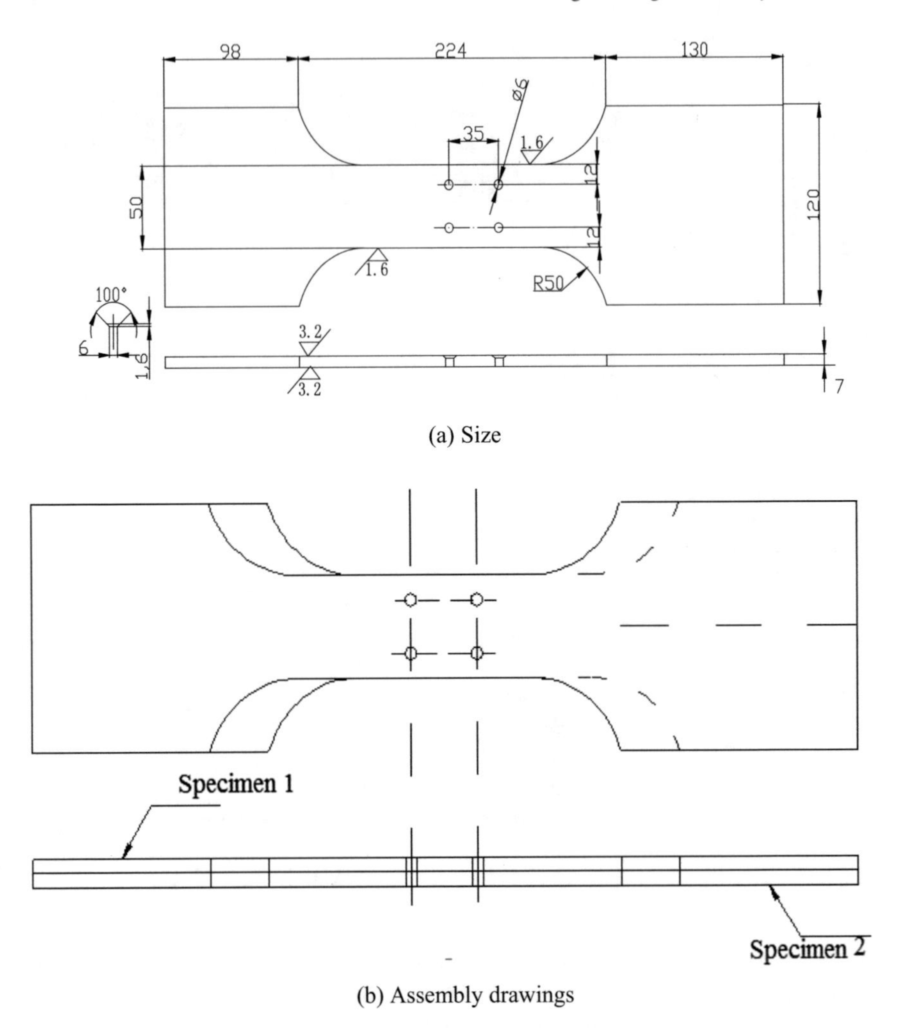

(a) Size

(b) Assembly drawings

Fig. 6.23 Double dogbone specimen

6.3.1 Nail Load Test

In order to study the effect of the material on the nail load of the reverse canine dog joint, the double dog bite test was carried out with different materials. The specimen material is imported aviation aluminum alloy 2024 and 7050-T7351, and the plate has a nominal thickness of 7 mm. Table 6.9 shows the connection parameters of the reverse double canine.

The position and serial number of the attached strain gauge are shown in Fig. 6.24.

Table 6.9 Important parameters of the reverse double dogbone specimen

Name	Material	Specimen width (mm)	Specimen thickness (mm)	Standard parts
100° countersunk head screw connector A	7050-T7351	50	14	KT-20S11-4/5
100° countersunk head screw connector B	2A12-T4	50	14	KT-20S11-4/5

A positive

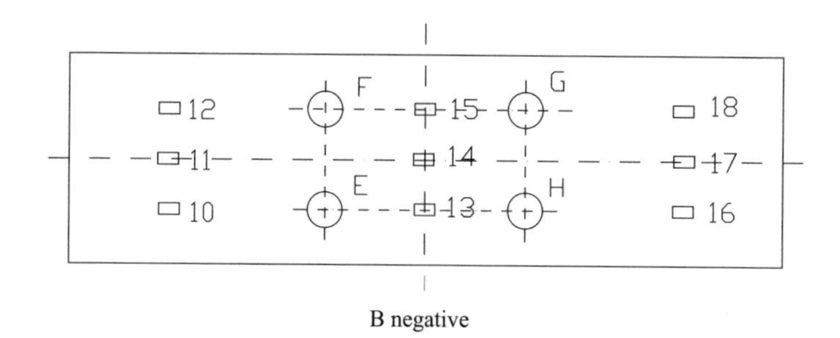

B negative

Fig. 6.24 Distribution of strain gauges on the specimen

The test load test of the test piece was completed on the INSTRON8802 hydraulic servo testing machine. The test piece is installed on the test machine as shown in Fig. 6.25. Load mode is load-time load, load rate: 1000 N/min. Each time you load 1 min, the data is collected once.

In Fig. 6.24, th strain obtained from points number 1, 2, 3.... 18 is ε_1, ε_2, ε_3 … ε_{18}. Pin A, B, C, D correspond to Pin E, F, G, H, respectively. It means the same pin has different number in two plates. We defined follows:

In Fig. 6.24 (A), the ratio of pin load A and B to plate load is R_1/P_1;
In Fig. 6.24 (A), the ratio of pin load C and D to plate load is R_2/P_2;
In Fig. 6.24 (B), the ratio of pin load E and F to plate load is R_3/P_3;
In Fig. 6.24 (B), the ratio of pin load G and H to plate load is R_4/P_4;

Fig. 6.25 Pin load test

There are

$$\frac{R_1}{P_1} = \left|\frac{(\varepsilon_1 + \varepsilon_2 + \varepsilon_3 - \varepsilon_4 - \varepsilon_5 - \varepsilon_6)}{(\varepsilon_1 + \varepsilon_2 + \varepsilon_3)}\right| \tag{6.10}$$

$$\frac{R_3}{P_2} = \left|\frac{(\varepsilon_7 + \varepsilon_8 + \varepsilon_9 - \varepsilon_4 - \varepsilon_5 - \varepsilon_6)}{(\varepsilon_7 + \varepsilon_8 + \varepsilon_9)}\right| \tag{6.11}$$

$$\frac{R_3}{P_3} = \left|\frac{(\varepsilon_{10} + \varepsilon_{11} + \varepsilon_{12} - \varepsilon_{13} - \varepsilon_{14} - \varepsilon_{15})}{(\varepsilon_{10} + \varepsilon_{11} + \varepsilon_{12})}\right| \tag{6.12}$$

$$\frac{R_4}{P_4} = \left|\frac{(\varepsilon_{16} + \varepsilon_{17} + \varepsilon_{18} - \varepsilon_{13} - \varepsilon_{14} - \varepsilon_{15})}{(\varepsilon_{16} + \varepsilon_{17} + \varepsilon_{18})}\right| \tag{6.13}$$

P_1 is the external load of the left end of the upper plate, P_2 is the external load of the right end of the upper plate; P_3 is the outer load of the lower end of the lower plate, and P_4 is the external load on the right end of the lower plate. According to the external load balance relationship, there are

$$P_1 + P_3 = P_2 + P_4 \tag{6.14}$$

The reverse biopsy of the canine bones was shown in Table 6.10.

Table 6.10 Distribution ration of rows (7050 and 2024)

	R_1/P_1 (%)	R_2/P_2 (%)	R_3/P_3 (%)	R_4/P_4 (%)
7050-T7351	2.8	7.3	18.8	2.4
2024	3.6	6.2	5.2	4.9

As can be seen from Table 6.10, for the 7050-T7351 material, the nail transmission ratio is very different, the maximum and minimum difference is of about 6 times, and for the 2A12-T4, the nail transmission ratio is compared close to the maximum and the smallest difference is less than 1 time. It can be seen that, due to the different materials, in the same connection mode, the nail transmission ratio is very different.

6.3.2 *Fatigue Test*

The fatigue test of the reverse double dog bone screw was done on the INSTRON8802 hydraulic servo testing machine. The stress ratio $R = 0.06$ and the loading frequency $f = 8$ Hz are divided into three kinds of stress levels: 130 MPa (low), 150 MPa (middle), and 170 MPa (high), respectively. Each group of stress level has 7 pieces. The test was carried out at room temperature.

After fatigue test, the mean value of life, variance, and coefficient of variation of the double canine bone screw pieces with different materials are shown in Table 6.11.

From Table 6.11, it can be seen that the life mean and the corresponding life dispersibility (coefficient of variation) of the 2024 reverse double dogbone joints are greater than 7050 reverse double dogbone joints and the corresponding life at different stress levels of data dispersibility. With the increase of the stress level, the fatigue life of the two kinds of materials was reduced, but the data dispersibility increased first and then decreased. That is, the lower the stress level is, the greater the dispersion of fatigue life, but the existence of a stress value makes the double dog bone screw fatigue life to the maximum dispersion. When the stress is far from this value, its fatigue life dispersion will be reduced.

After fatigue test, all fractures are observed by stereomicroscope. For reverse double dongbone specimens made of 7050 and 2024 plates, it can be seen that crack always initiated at the hole edge where two plates contact each other. Crack is in the form of corner crack, as shown in Fig. 6.26. It can be seen from Fig. 6.26 that the crack propagation velocity in the depth direction is smaller than the crack propagation velocity in the longitudinal direction, and the crack surface is oval, which is caused by the introduction of compressive stress in the thickness direction of the plate due to pre-tension load of the bolt or rivet.

Table 6.11 Data processing of double dogbone screw specimen

Stress level (MPa)	7050-T351		2024	
	Life cycle (cycle)	Coefficient of variation	Life cycle (cycle)	Coefficient of variation
130	120,522	0.23692	227467	0.38608
150	60,827.9	0.54611	118,686	1.0038
170	44,820.2	0.30311	81,895	0.7323

Fig. 6.26 Macro cross-section of double dog bone pieces

6.3.3 Maximum Principal Stress Analysis

Taking aluminum alloy 2024 as an example, ABAQUS finite element software was used to establish the three-dimensional elastoplastic finite element model of reverse double canine bone and was calculated. The Young's modulus $E = 73$ GPa, the yield stress $\sigma_y = 322$ MPa, and the Poisson's ratio $v = 0.33$ were determined by the experiment. The linear hexahedral reduction integral unit C3D8R is used for meshing. In order to improve the accuracy of the calculation, the perimeter mesh is further refined, see Fig. 6.27.

Three kinds of finite element models are considered, one end of the model is fixed and the other end is applied with 130, 150, and 170 MPa tensile load respectively. The bolt preload is set to 6000 N.

In the finite element model, the following four contacts are defined:

(A) The contact between the upper and lower plates,
(B) The contact between the bolt and the wall,
(C) The contact between the bolt head and the upper plate,
(D) The contact between the nut and the upper plate.

The friction effect was included in this work by using Elastic Coulomb friction with $\mu = 0.3$.

The test results show that the crack always starts at the hole of the upper and lower plates, see Fig. 6.26. The finite element calculation results show that the maximum principal stress of the reverse double dogbone joints also occurs at that position, see Fig. 6.28.

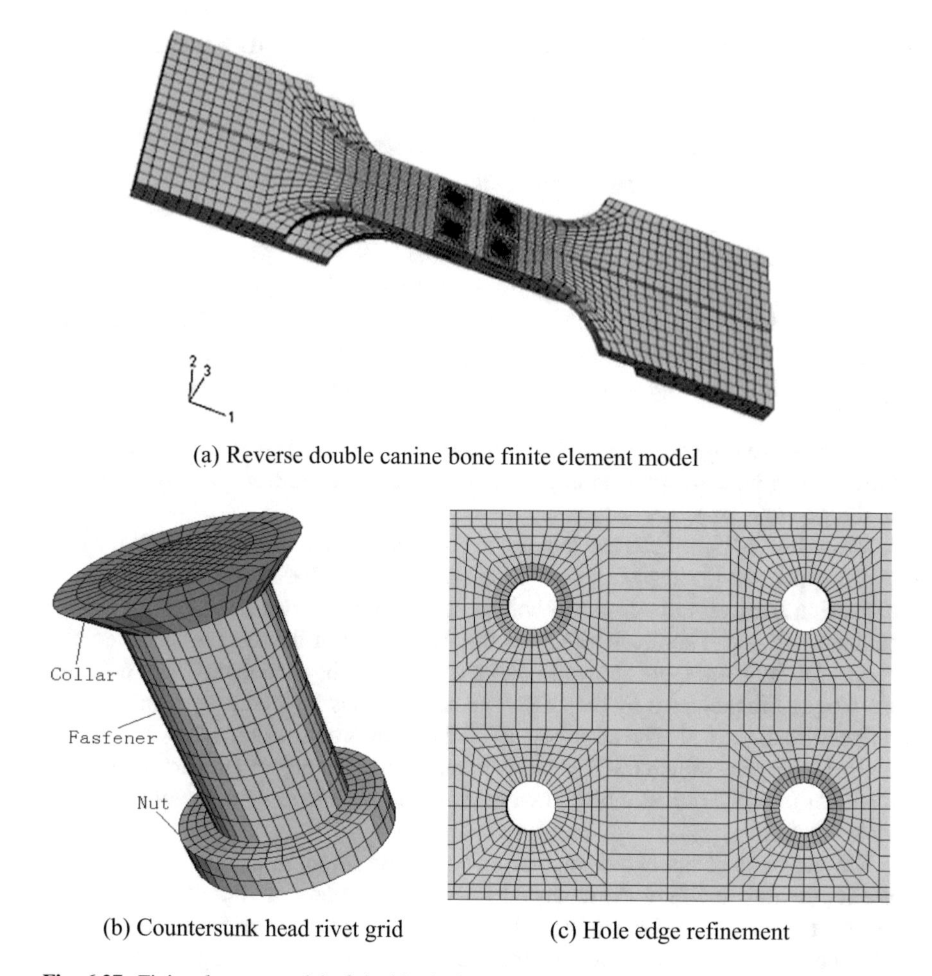

(a) Reverse double canine bone finite element model

(b) Countersunk head rivet grid

(c) Hole edge refinement

Fig. 6.27 Finite element model of double dogbone specimen

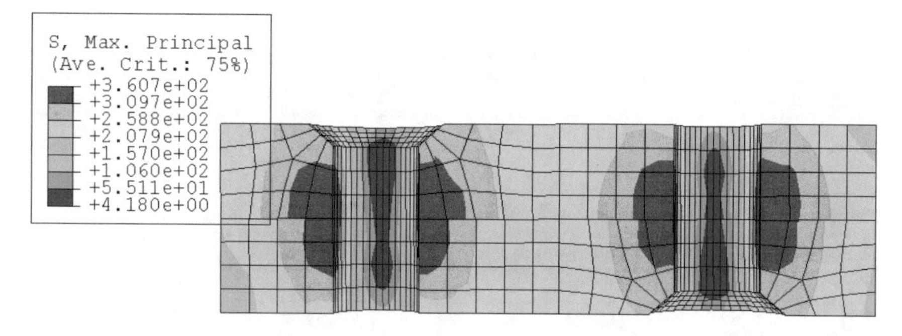

Fig. 6.28 Reverse double dog bone countersunk head fastening hole stress contour

6.4 Application of Multi-axis Fatigue Theory on Fatigue Life Prediction of Joints

6.4.1 *Life Prediction of Aluminum Alloy Reverse Double Dogbone Joints Specimen*

Based on the finite element calculation results in Sect. 6.3.3, we use the Smith–Watson–Topper (SWT) and Wang–Brown (WB) models based on the critical plane method in Sect. 2.4.3 for the 2024 reverse double dogbone hole fatigue life prediction and the predicted results are compared with the experimental results. The fatigue performance parameters of the 2024 aluminum alloy are shown in Table 6.12.

At the same time, in order to study the effect of bolt preload on the fatigue life of the fastening hole, the preload is set to 5000, 7000, 8000, and 9000 N, respectively, in the previous model.

From Fig. 6.28, we can see that the maximum principal stress occurs at the hole of the upper and lower plates, and the test results show that the crack initiation and expansion here, so the stress can be used to predict the fatigue life of the double canine specimen. The mean values of the fatigue life of the test and the predicted values for fatigue life are shown in Table 6.13. It can be seen from Table 6.13 that the fatigue life predicted by the critical plane method is in good agreement with the fatigue life of the test. The accuracy of SWT model is better than that of WB model under high-cycle fatigue (HCF), and the prediction accuracy of WB model is lower than that of SWT model under low-cycle fatigue (LCF) for reverse double canine fastening hole.

On the basis of the above analysis, we can use the multi-axis fatigue theory to predict the fatigue life of the reverse double canine under different preload. Fatigue life prediction was performed using SWT model when the remote tensile load was 130 and 150 MPa (high-cycle fatigue). Fatigue life prediction was performed using the WB model when the farfield tensile load was 170 MPa (low-cycle fatigue). The predicted values of the fatigue life of the connecting members under different

Table 6.12 Fatigue properties of aluminum alloy 2024 [7]

Fatigue parameters	b	c	σ_f'(MPa)	ε_f'	s	υ_p
	−0.124	−0.65	1103	0.22	0.3	0.5

Table 6.13 Fatigue life of fastening holes in reverse double canine bone specimen

Stress level/MPa	Test results/cycle	SWT/cycle	WB/cycle
130	227,467	393,805	956,807
150	118,686	116,403	626,295
170	81,895	28,191	95,236

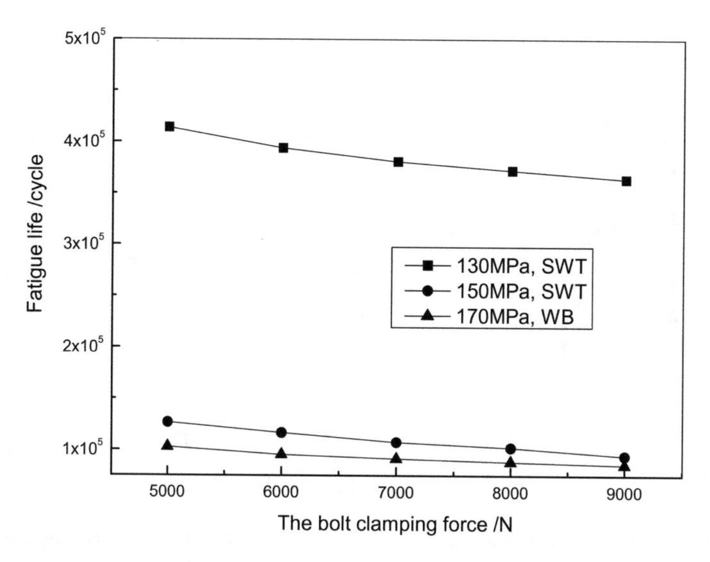

Fig. 6.29 Fatigue life under different preloads

preload forces are shown in Fig. 6.29. It can be seen that the fatigue life of the fastening holes decreases with the increase of the pretensioning force.

6.4.2 Estimation of Life of Single Shear Lap Joints

The prediction of the multi-axial fatigue life of the scarfed lap structure in Sect. 6.1 is carried out. Based on the SWT model and the WB model, the fatigue life of the slanting fastening hole is predicted based on the finite element calculation results. As can be seen from Fig. 6.30, the WB model predicts a lower life expectancy, tends to be conservative, and tends to cause design waste. The SWT model predicts a slightly larger life, but is closer to the real life. This is consistent with the predicted results of the fatigue life of the fastening hole of the double canine bone specimen, that is, the SWT model is suitable for the prediction of high-cycle fatigue life, and the WB model is more suitable for the prediction of low-cycle fatigue life.

The fatigue life of the scarfed lap structure is compared with the predicted life of the SWT multi-axis fatigue theory, which is plotted in Fig. 6.31. It can be seen that most data points fall within the dispersion band with the error factor of 2, indicating that the life prediction results of the SWT model are reliable.

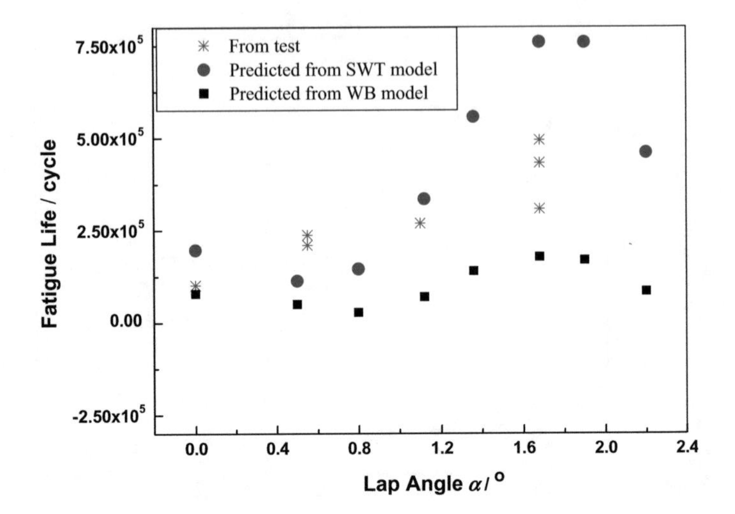

Fig. 6.30 The effect of the overlapping angle of the scarfed lap structure on the fatigue life of the fastening hole

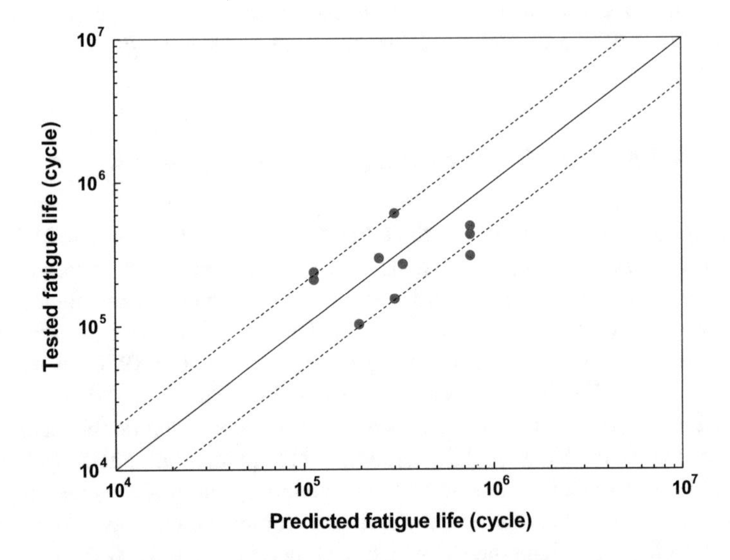

Fig. 6.31 Comparison of tested life and SWT predictive fatigue life of lap joints

References

1. Cheng, Baoqu. 1990. *Aircraft Construction Process*, 4. Beijing: National Defense Industry Press.
2. (Aircraft Design Manual) Editorial Board. 2001. *Aircraft Design Handbook: Load, Strength and Stiffness*. Beijing: Aviation Industry Press.

3. Liao, M., G. Shi, and Y. Xiong. 2001. Analytical Methodology for Predicting Fatigue Life Distribution of Fuselage Splices. *International Journal of Fatigue* 23: S177–S185.
4. Chakherlou, T.N., R.H. Oskouei, and J. Vogwell. 2008. Experimental and Numerical Investigation of the Effect of Clamping Force on the Fatigue Behaviour of Bolted Plates. *Engineering Failure Analysis* 15 (5): 563–574.
5. Wu, Fuming. 1985. *Structural Fatigue Strength*. Xi'an: Northwestern Polytechnical University Press.
6. Park, Chul Young, and Alten F. Grandt Jt.2007. Effect of Load Transfer on the Cracking Behavior at a Countersunk Fastener Hole. *International Journal of Fatigue* 29: 146–157.
7. Wang, Zhongguang. 1998. *Material Fatigue*, 182–184. Beijing: National Defense Industry Press.

Chapter 7
Fatigue Test and Analysis of Box Section

Due to requirements for weight reduction and structural safety, thin-walled structure is widely used in the aircraft, such as stiffened panels, spars, and box sections composed by skin, beams, and ribs. The most common failure mode of thin-walled structures is strength failure [1, 2]. In order to ensure the safety of the aircraft, the corresponding strength check is needed. The component-level strength test of the building block system was carried out on the box section. The purpose of this study is to evaluate the correlation analysis method, and to provide support for the validation test of the aircraft structure.

7.1 Box Section Fatigue Test

As the key part of the damage tolerance design [3, 4], the lower aerofoil skin of the wing box section is in tensile stress state for most cases during the actual flight. The purpose of this test is to determine the fatigue life of a wing box section model and verify the accuracy of the deterministic crack growth method by comparing with the experimental results.

7.1.1 Specimen

The specimens for this fatigue test are wing box section model of an airplane, which consists of 2 outer skins, 2 outer plates, 1 rib, 16 beam edge strips, 4 reinforced-ribbed strips, 4 webs, 1 conversion part, 1 set of fixtures, multiple angle plate, and reinforced profiles. The layouts and locations of each component are shown in Fig. 7.1. The box section is shown in Fig. 7.2.

J. Liu et al., *Long-Life Design and Test Technology of Typical Aircraft Structures*,
https://doi.org/10.1007/978-981-10-8399-0_7

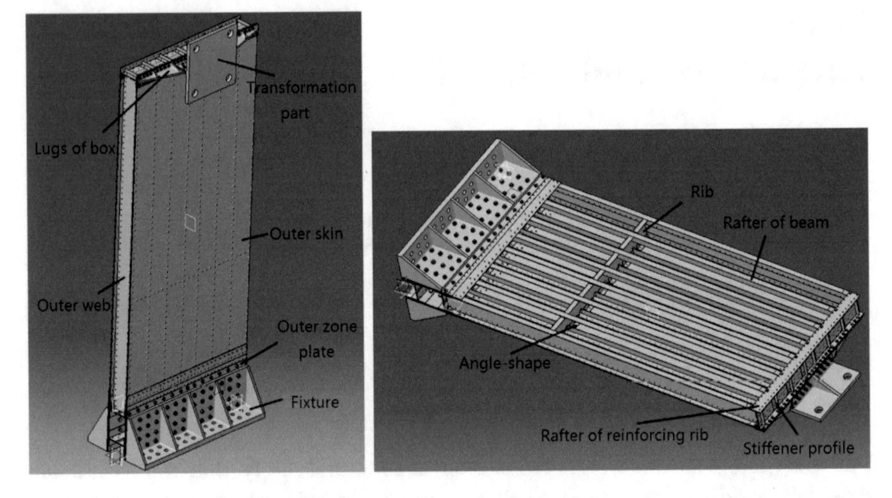

Fig. 7.1 Component layout and location

Fig. 7.2 Box section model and object

7.1.2 Loading Method and Test Results

The installation and loading scheme of the box section is shown in Fig. 7.3. The bracket is fixed on the bearing wall by bolting. The loading cylinder is fixed on the bracket by bolting. The base is fixed on the ground by bolting. The root of box section is fixed on the horizontal base through bolting. In order to realize loading, the transition part of the free end of the box section is connected with the loading end of the oil cylinder through bolting. In this loading scheme, the box section is installed vertically, which can effectively weaken the influence of the weight of the box on the fatigue test results.

As shown in Fig. 7.4, in order to ensure the loading mode does not limit the displacement of the free end of the box section in the vertical direction and avoid the additional load on the section web, U-shaped lug goes through the bolt hole of the lug. U-shaped lug is connected to the two lugs by bolting. The two lugs are connected to the loading end of the oil cylinder through a combined steel sheet.

The fatigue tests employ "FTS- complex loading system", and the fatigue load spectrum of box section is a sinusoidal amplitude spectrum with $F_{max} = 20$ KN, stress ratio of 0 and the loading frequency of 0.2 Hz. The load direction is perpendicular to the skin. During the test, the surface and the edge of each box were observed regularly.

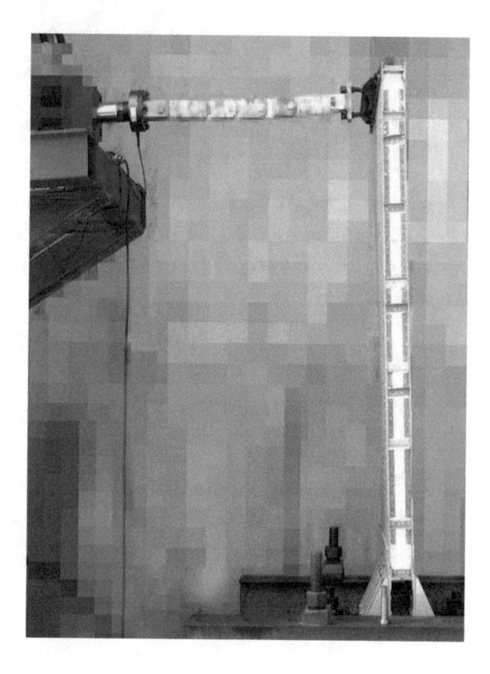

Fig. 7.3 Box section installation and loading scheme

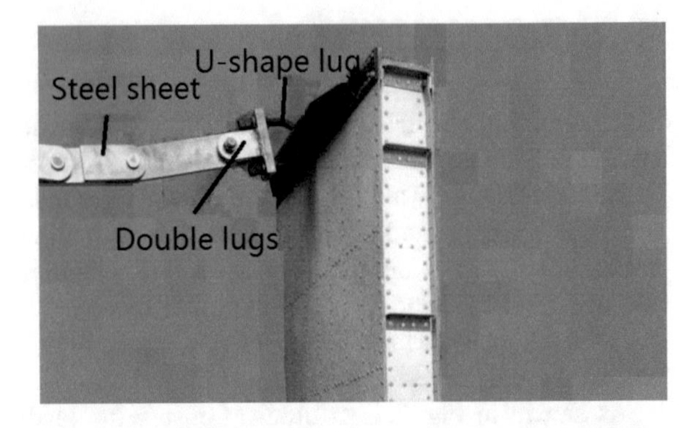

Fig. 7.4 Box section-loaded connection mode

Fig. 7.5 Box section skin failure

After a total of 52,673 cycles of fatigue test, more than 15 mm of penetrating cracks appeared on the skin of the box section, which reached the end of the fatigue test. As shown in Fig. 7.5, the penetrating cracks appear on the outer skin of the tension side.

7.2 Box Section Life Evaluation

In this section, the life prediction of the box section model specimen is carried out by the deterministic crack growth analysis (DCGA) method based on fracture mechanics and crack growth rate model [5, 6].

7.2.1 Box Section Model and Results

The Catia model was imported into the finite element software ABAQUS as the relevant box model. Due to the symmetry of box structure and load, only half of the model was taken to calculate (Fig. 7.6). In order to facilitate meshing and computing, most of the bolts were replaced by reference points. Through analysis, the skin and the beam edge transfer the load to the strip plate and the reinforcing rib strip at the root of the box section. The strip plate and the reinforcing rib edge strip are the regions with higher stress level, so the bolts and rivets on them are not simplified.

For the finite element model of the wing box section, it is necessary to provide test parameters and compare with those of test. Therefore, the load and boundary conditions from finite element model should be in accord with that from the test. In the finite element model, the displacements in *X*, *Y*, *Z* directions of the bottom plate of the transition plate were constrained, and the symmetric boundary conditions were applied on the symmetry plane. The tensile load is half of the total load that is 10 kN.

According to the results of finite element analysis, comparing the stress level of the high-stress zone of the main components, the hole edge stress under the skin is maximum, and the stress near the lateral web under the skin and beam flange connection hole side is also large (see Fig. 7.7).

After comprehensive consideration, the hole under the skin is the most dangerous nail hole of the box section model due to its maximum Mises stress. The connecting hole of the lower skin near the outer web and the edge of the beam can be considered as a secondary danger hole. The two holes mentioned above are shown in Fig. 7.8, and they are used to evaluate the fatigue life.

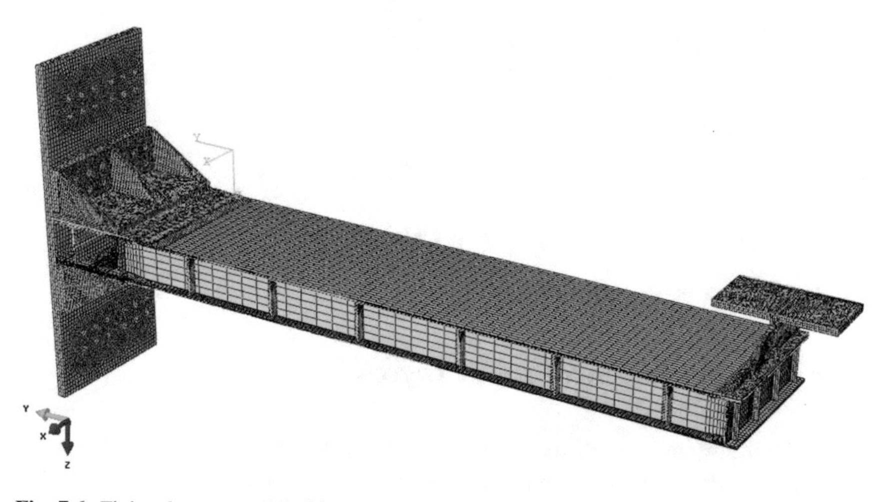

Fig. 7.6 Finite element model of box section

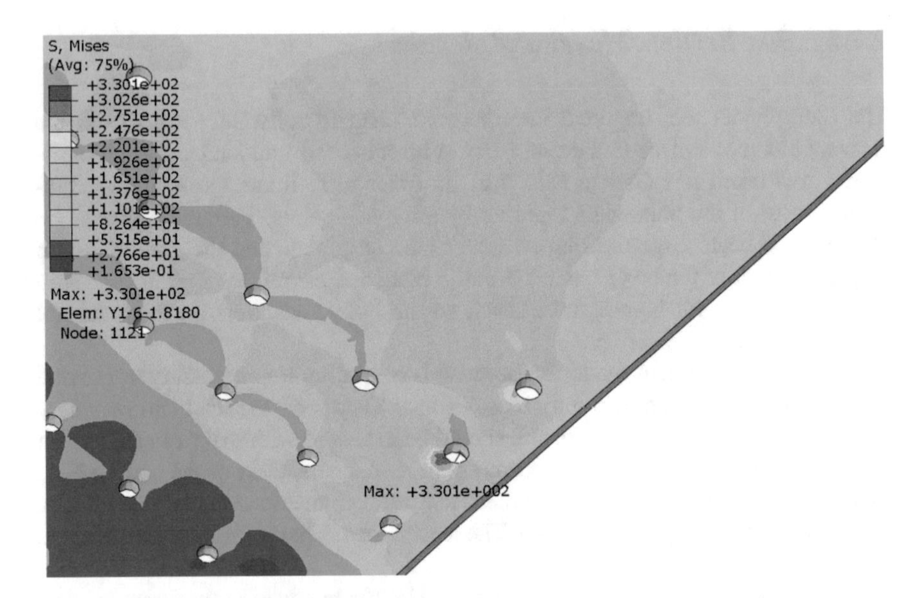

Fig. 7.7 Mises stress under the skin

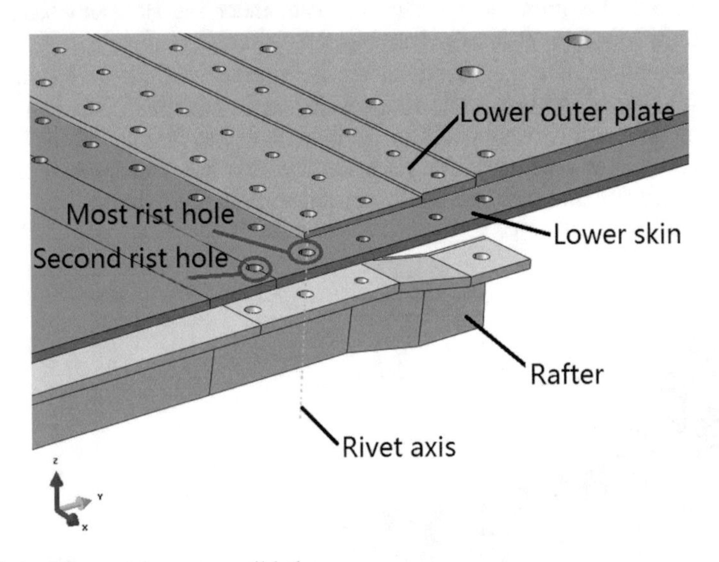

Fig. 7.8 Locations of dangerous nail holes

7.2.2 *Shape and Size of Initial Defects*

In the durability analysis of the structure with the durability of the crack growth method, it is necessary to obtain the test data with the same original fatigue quality. The specimen used in the original fatigue quality evaluation is a single dog bone

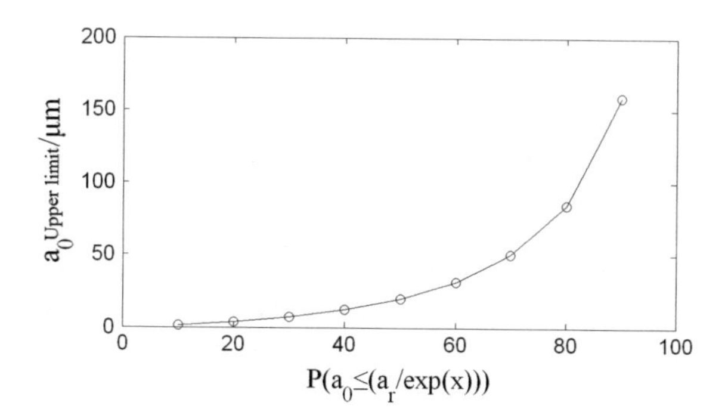

Fig. 7.9 Upper bound a_0 and probability P relation curve

aluminum alloy riveted structure specimen, which is the same as the outer skin material, thickness, manufacturing process, and rivet. The most dangerous nail holes and the secondary dangerous nail holes are under the skin, riveting the same way, so their original fatigue quality is the same. According to the requirement of durability analysis, it is assumed that the shape of the initial defect is a single corner crack of quarter-circular arc hole. The probability distribution curve of the initial flaw size of the dangerous nail hole has been determined by the previous experiment. The relationship between the upper bound of the initial crack size a_0 and the probability P is shown in Fig. 7.9.

As shown in Fig. 7.9, when the probability P increases, the upper bound of the initial crack size a_0 increases, and different dimensions will directly lead to different life under the damage tolerance framework. Taking different probability P ranging from 10 to 90%, the upper bound of the initial crack size a_0 is calculated as the starting point of crack propagation (see Table 7.1).

7.2.3 *Fatigue Crack Propagation Analysis Program*

The load spectrum of the box section fatigue test load is a sine wave with maximum load of 20 KN, stress ratio of $R = 0$. The Paris equation is used to describe the crack propagation

$$\frac{da}{dN} = C(\Delta K)^n \tag{6.1}$$

Table 7.1 The upper bound of initial crack size a_0 under a different probability P

Probability P	10%	20%	30%	40%	50%
a_0 (mm)	0.0016	0.0038	0.0073	0.0123	0.0199
Probability P	60%	70%	80%	90%	
a_0 (mm)	0.0316	0.0505	0.0844	0.1584	

in which C and n are materials parameters determined by experiment, depending on materials and stress ratio.

The most dangerous rivet was taken out to investigate. The determination of the stress intensity factor at the crack tip of the rivet hole can be illustrated as shown in Fig. 7.10 (Case A). The rivet hole is partly affected by the load F, and the two ends are subjected to uniform tensile stress σ_1 and σ_2, respectively. It can be well proved [7]: the stress intensity factor of Case A is equivalent to the superposition of the stress intensity factor of Case B (the two ends are subjected to uniform tensile stress $(\sigma_1 + \sigma_2)/2$) and the half stress intensity factor of Case C (the rivet hole is partly affected by the load F). It can be written as

$$K_A = K_B + \frac{1}{2} K_C$$

in which K_A is the stress intensity factor of Case A, K_B is the stress intensity factor of Case B, K_C is the stress intensity factor of Case C. The key of solving the stress intensity factor K_A of Case A is to find out the uniform tensile stress σ_1 and σ_2 and concentrated load F.

Based on the finite element analysis and the equivalent SSF method [8], the calculation results of the stress intensity factor at the crack tip of the most and the secondary dangerous rivet holes can be obtained (see Table 7.2).

For fatigue dangerous regions, the materials used for the lower skin is 2A12-T4. According to the local size and stress ratio of the rivet, the material parameters of the crack propagation are obtained from the manual [9] [7]:

$$C = 2.008 \times 10^{-8} \text{mm}/\left(\text{cycle} \cdot (\text{N/m})^{2/3}\right)^{3.67}, \quad n = 3.67.$$

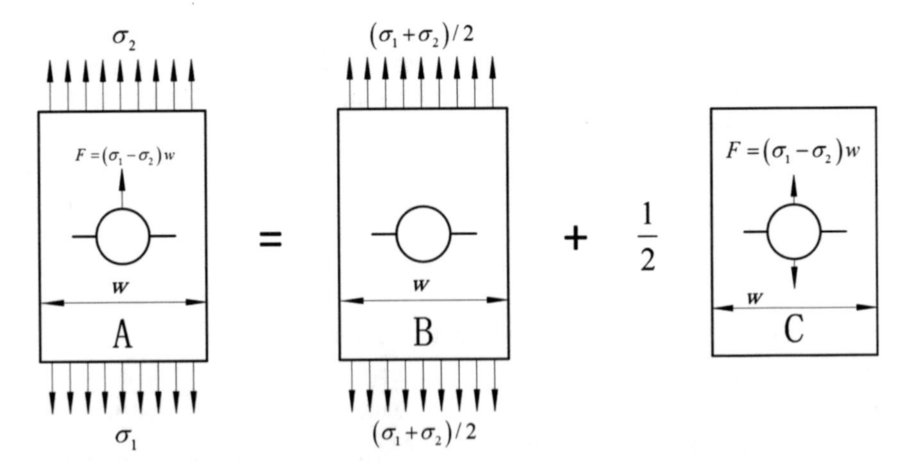

Fig. 7.10 Schematic view of crack tip stress intensity factor decomposition

Table 7.2 Calculated stress intensity factor of crack tip of rivet holes by finite element method

Dangerous parts	σ_1/ MPa	σ_2/ MPa	F/ N	w/ mm
The most dangerous hole	181.1	162.6	740	20
Times dangerous hole	172.3	153.4	710	20

Adopting the Paris crack growth rate model, the fracture toughness of 2A12-T4 is calculated as the end point of crack growth. In the framework of damage tolerance, the fatigue life of the dangerous parts of the aircraft is calculated. The fracture toughness of 2A12-T4 is found to be between 33 and 37 MPa$\sqrt{\text{m}}$ in general from manual [10]. For the sake of conservatism, this section takes its minimum value of 33 MPa$\sqrt{\text{m}}$.

The analysis flow of crack propagation is shown in Fig. 7.11.

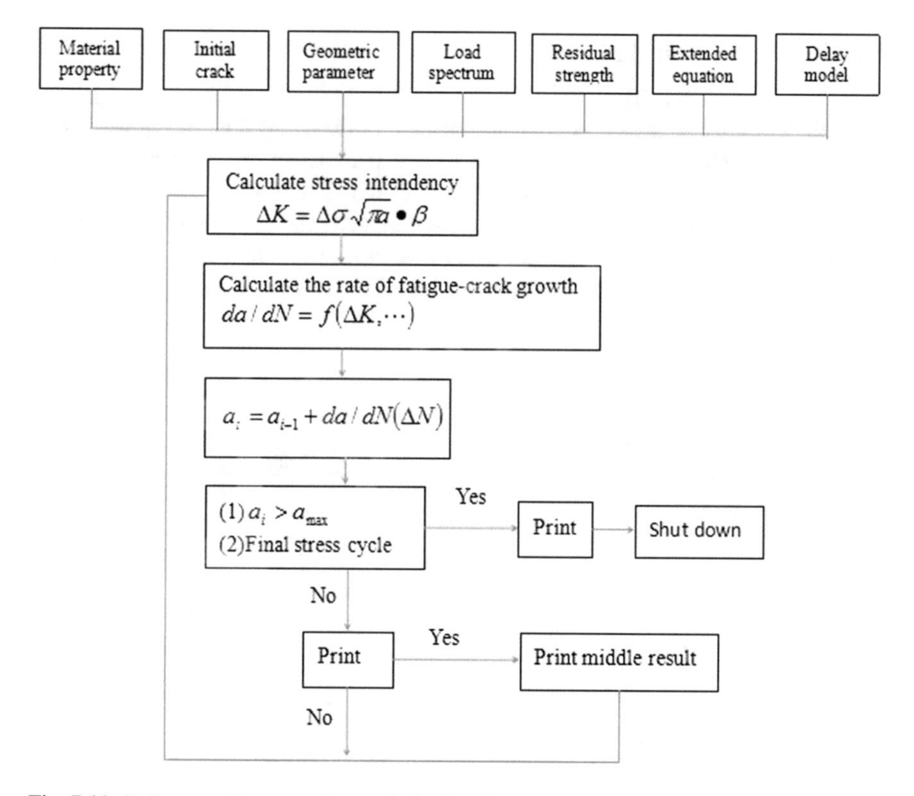

Fig. 7.11 Fatigue crack propagation analysis flowchart

Table 7.3 Evaluation results of box life under a different probability P of the most dangerous hole

Probability P	10%	20%	30%	40%	50%
Life N/cycles	68,594	55,206	48,253	41,203	35,128
Probability P	60%	70%	80%	90%	
Life N/cycles	33,251	30,126	28,674	26,342	

Table 7.4 Evaluation results of box life under different probability P of the secondary dangerous holes

Probability P	10%	20%	30%	40%	50%
Life N/cycles	92,135	68,453	54,237	48,253	43,231
Probability P	60%	70%	80%	90%	
Life N/cycles	37,125	34,157	33,251	30,674	

7.2.4 Evaluation Results

The initial crack size a_0, stress intensity factor K_A of Case A, the crack propagation material parameters C and n and other related parameters are used by the crack growth calculation program compiled by MATLAB. The results of box section life evaluation under a different probability P are obtained. The life evaluation results of the most dangerous hole and the secondary dangerous hole are shown in Tables 7.3 and 7.4. The probability P and the box section life curve are obtained as shown in Figs. 7.12 and 7.13.

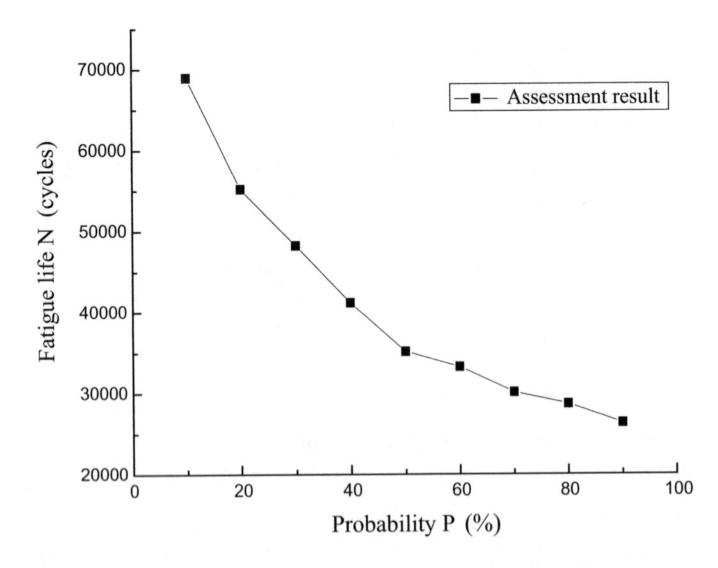

Fig. 7.12 The most dangerous hole nail probability P and box section life curve

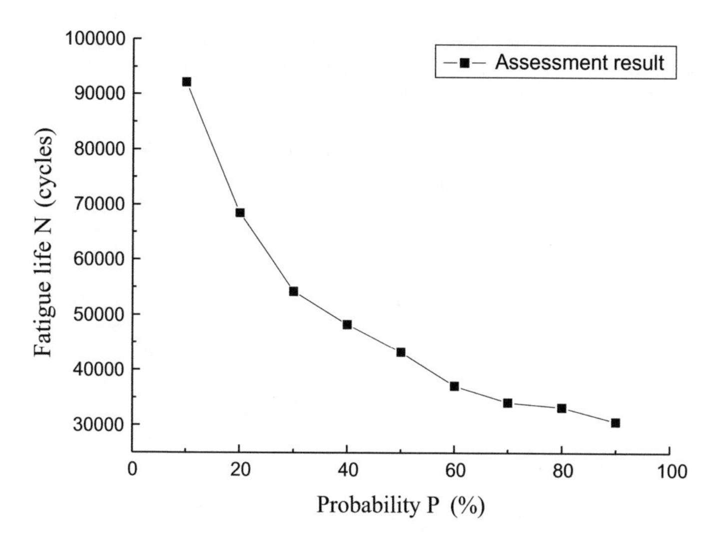

Fig. 7.13 The secondary dangerous hole nail probability *P* and box section life curve

With the increase of the probability of *P*, the upper bound of the initial crack size a_0 also increased, and the box section life evaluation results obtained by the deterministic crack growth method are decreased. It is shown that the larger the initial crack size is, the shorter the life of the box section will be. The life of the box section is very sensitive to the initial crack size. In the actual structure, the original fatigue quality of the hole can be improved through the improvement of the drilling process and the rivet quality, thereby the purpose of improving the fatigue life of the whole structure is achieved.

References

1. Tao, Meizheng, Qin Sun, and Jianliang Ai. 2001. *Comprehensive Design of Modern Aircraft Structure*. Xi'an: Northwestern Polytechnical University Press.
2. Lu, Binghe, Xiaopeng Wan, and Meiying Zhao. 2009. Stability Analysis of Composite Box Section Structure. *China Mechanical Engineering* 20 (17): 2055–2058.
3. Ye, Tianqi, and Tianxiao Zhou. 1996. *A Guide to Finite Element Analysis of Aeronautical Structures*. Beijing: Aviation Industry Press.
4. Li, Shanshan, Lingge Xing, Huiliang Ding, and Shihui Deng. 2011. Study on Optimization Method of Damage Tolerance in Initial Design Optimization of Aircraft Structures. *Advances in Aviation Engineering* 2 (1): 78–83.
5. China Aviation Research Institute. 1993. Stress Strength Factor Manual. Beijing: Science Press.
6. Zhang, Chengcheng, Weixing Yao, and Bing Ye. 2009. Equivalent SSF Method for Fatigue Life Analysis of Joints. *Journal of Aeronautical Engineering* 2: 271–275.
7. Xueren, Wu. 1996. *Aircraft Structure Metal Material Mechanical Performance Manual (Volume 1)*. Beijing: Aviation Industry Press.

8. Liu, Wenting, and Minzhong Zheng. 1999. *Probabilistic Fracture Mechanics and Probability Damage Tolerance/Durability*. Beijing: Beijing Aerospace Publishing House.
9. Dong, Yanming, and Wenting Liu. 2010. Generic Analysis and Experimental Study on Eifs Distribution. *Mechanical Strength* 32 (5): 795–800.
10. Aircraft Design Handbook. 2001. Book 3. Beijing: Aviation Industry Press.

Printed in the United States
By Bookmasters